A-Z
SEMICONDUCTOR

A-Z SEMICONDUCTOR

Dr. Alimuddin Khan

CENTRUM PRESS
NEW DELHI-110002 (INDIA)

CENTRUM PRESS
H.O.: 4360/4, Ansari Road, Daryaganj,
New Delhi-110 002 (India)
Ph.: 23278000, 23261597
B.O.: No. 1015, Ist Main Road, BSK IIIrd Stage
IIIrd Phase, IIIrd Block,
Bangalore - 560 085 (India)
Tel.: 080-41723429
Visit us at: www.centrumpress.com

A-Z Semiconductor

First Edition, 2009
ISBN 978-93-80106-36-6

PRINTED IN INDIA

Printed at Salasar Imaging Systems, Delhi-110035 (India)

Contents

Preface

A-Z Semicanductor is motivated by the experience of teaching the Physics course in the Electronics Engineering program at the University of Amravati. This book is designed for students, engineers, and scientists seeking a comprehensive background in Modern physics in the minimum time with the minimum number of mathematical and physical concepts.

In this book author have provided a very extensive and detailed text on the physics underlying semiconductor devices. More so than any other current text, this book provides a great expanded discussion of modern tight-binding methods, helping the students to understand these aspects of electronic structure in clear, simple terms.

In connection with this the author offer a very detailed discussion of deep levels in semiconductors, which are so important to semiconducting properties. Also, in the discussion of transport properties, the book goes into much greater depth about nonlinear and nonequilibrium processes than is usual. It is quite a unique contribution, containing the basic physics which tends to be missing from device-oriented books, but going much further into the essentials needed for device development.

Author

Chapter 1

Amplifiers and Active Devices

FROM ELECTRIC TO ELECTRONIC

The transition between *electric* circuits and *electronic* circuits is formally crossed. Electric circuits are connections of conductive wires and other devices whereby the uniform flow of electrons occurs.

Electronic circuits add a new dimension to electric circuits in that some means of *control* is exerted over the flow of electrons by another electrical signal, either a voltage or a current.

In and of itself, the control of electron flow is nothing new to the student of electric circuits. Switches control the flow of electrons, as do potentiometers, especially when connected as variable resistors (rheostats). Neither the switch nor the potentiometer should be new to your experience by this point in your study.

The threshold marking the transition from electric to electronic, then, is defined by *how* the flow of electrons is controlled rather than whether or not any form of control exists in a circuit.

Switches and rheostats control the flow of electrons according to the positioning of a mechanical device, which is actuated by some physical force external to the circuit. In electronics, however, we are dealing with special devices able to control the flow of electrons according to another flow of electrons, or by the application of a static voltage. In other words, in an electronic circuit, *electricity is able to control electricity*.

The historic precursor to the modern electronics era was invented by Thomas Edison in 1880 while developing the electric incandescent lamp. Edison found that a small current passed from the heated lamp filament to a metal plate mounted inside the vacuum envelop. Today this is known as the "Edison effect". Note that the battery is only necessary to heat the filament. Electrons would still flow if a non-electrical heat source was used.

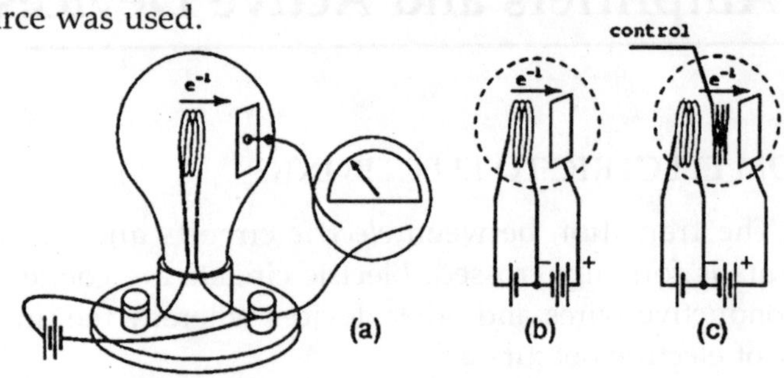

Fig. (a) Edison Effect, (b) Fleming Valve or Vacuum Diode, (c) De Forest Audion Triode Vacuum Tube Amplifier.

By 1904 Marconi Wireless Company adviser John Flemming found that an externally applied current (plate battery) only passed in one direction from filament to plate, but not the reverse direction (not shown). This invention was the vacuum diode, used to convert alternating currents to DC. The addition of a third electrode by Lee DeForest allowed a small signal to control the larger electron flow from filament to plate.

Historically, the era of electronics began with the invention of the *Audion tube,* a device controlling the flow of an electron stream through a vacuum by the application of a small voltage between two metal structures within the tube.

Electronics technology experienced a revolution in 1948 with the invention of the *transistor*. This tiny device achieved approximately the same effect as the Audion tube, but in a vastly smaller amount of space and with less material. Transistors control the flow of electrons through solid *semiconductor* substances rather than through a vacuum, and so transistor technology is often referred to as *solid-state* electronics.

ACTIVE VERSUS PASSIVE DEVICES

An *active* device is any type of circuit component with the ability to electrically control electron flow (electricity controlling electricity). In order for a circuit to be properly called *electronic,* it must contain at least one active device. Components incapable of controlling current by means of another electrical signal are called *passive* devices. Resistors, capacitors, inductors, transformers, and even diodes are all considered passive devices. Active devices include, but are not limited to, vacuum tubes, transistors, silicon-controlied rectifiers (SCRs), and TRIACs. A case might be made for the saturable reactor to be defined as an active device, since it is able to control an AC current with a DC current, but I've never heard it referred to as such.

All active devices control the flow of electrons through them. Some active devices allow a voltage to control this current while other active devices allow another current to do the job.

Devices utilizing a static voltage as the controlling signal are, not surprisingly, called *voltage-controlled* devices. Devices working on the principle of one current controlling another current are known as *current-controlled* devices. For the record, vacuum tubes are voltage-controlled devices while transistors are made as either voltage-controlled or current controlled types. The first type of transistor successfully demonstrated was a current-controlled device.

AMPLIFIERS

The practical benefit of active devices is their *amplifying* ability. Whether the device in question be voltage-controlled or current-controlled, the amount of power required of the controlling signal is typically far less than the amount of power available in the controlled current. In other words, an active device doesn't just allow electricity to control electricity; it allows a *small* amount of electricity to control a *large* amount of electricity.

Because of this disparity between *controlling* and *controlled* powers, active devices may be employed to govern a large amount of power (controlled) by the application of a small amount of power (controlling). This behavior is known as *amplification*.

It is a fundamental rule of physics that energy can neither be created nor destroyed. Stated formally, this rule is known as the Law of Conservation of Energy, and no exceptions to it have been discovered to date.

If this Law is true — and an overwhelming mass of experimental data suggests that it is — then it is impossible to build a device capable of taking a small amount of energy and magically transforming it into a large amount of energy. All machines, electric and electronic circuits included, have an upper efficiency limit of 100 percent. At best, power out equals power in as in Figure below.

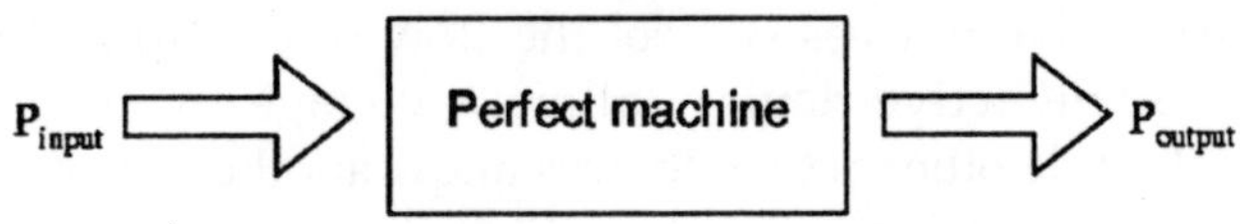

$$\text{Efficiency}-\frac{P_{output}}{P_{input}} < 1 = \text{less than } 100\%.$$

Fig. The Power Output of a Machine can Approach, but never exceed, the Power Input for 100% Efficiency as an Upper Limit.

Usually, machines fail even to meet this limit, losing some of their input energy in the form of heat which is radiated into surrounding space and therefore not part of the output energy stream.

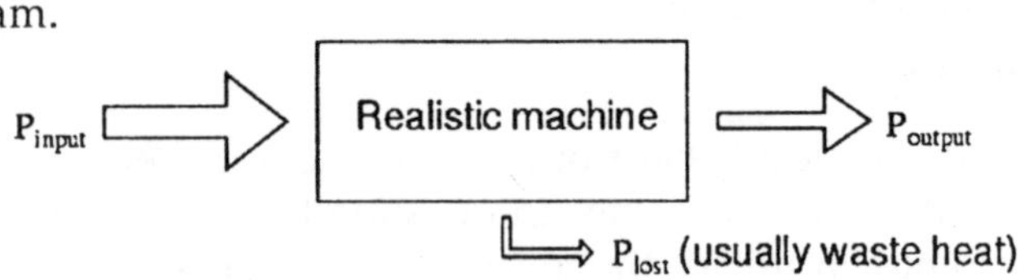

$$\text{Efficiency}-\frac{P_{output}}{P_{input}} < 1 = \text{more than } 100\%.$$

Fig. A Realistic Machine Most Often Loses Some of its Input Energy as Heat in Transforming it into the Output Energy Stream.

Many people have attempted, without success, to design and build machines that output more power than they take in. Not only would such a *perpetual motion* machine prove that the Law of Conservation of Energy was not a Law after all, but it would usher in a technological revolution such as the world has never seen, for it could power itself in a circular loop and generate excess power for "free".

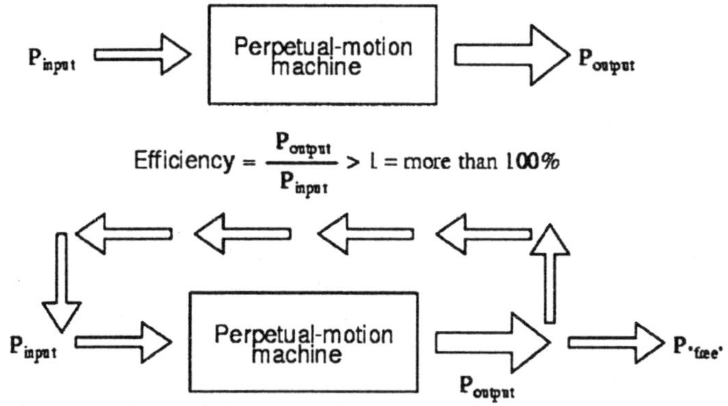

Fig. Hypothetical "Perpetual Motion Machine" Powers Itself?

Despite much effort and many unscrupulous claims of "free energy" or *over-unity* machines, not one has ever passed the simple test of powering itself with its own energy output and generating energy to spare.

There does exist, however, a class of machines known as *amplifiers*, which are able to take in small-power signals and output signals of much greater power. The key to understanding how amplifiers can exist without violating the Law of Conservation of Energy lies in the behavior of active devices.

Because active devices have the ability to *control* a large amount of electrical power with a small amount of electrical power, they may be arranged in circuit so as to duplicate the form of the input signal power from a larger amount of power supplied by an external power source.

The result is a device that appears to magically magnify the power of a small electrical signal (usually an AC voltage waveform) into an identically-shaped waveform of larger magnitude.

The Law of Conservation of Energy is not violated because the additional power is supplied by an external source, usually a DC battery or equivalent. The amplifier neither creates nor destroys energy, but merely reshapes it into the waveform desired.

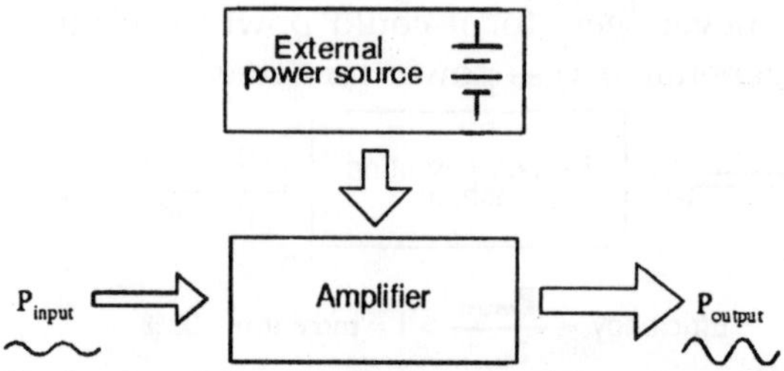

Fig. While An Amplifier can Scale a small Input Signal to Large Output, its Energy Source is an External Power Supply.

In other words, the current-controlling behavior of active devices is employed to *shape* DC power from the external power source into the same waveform as the input signal, producing an output signal of like shape but different (greater) power magnitude. The transistor or other active device within an amplifier merely forms a larger *copy* of the input signal waveform out of the "raw" DC power provided by a battery or other power source.

Amplifiers, like all machines, are limited in efficiency to a maximum of 100 percent. Usually, electronic amplifiers are far less efficient than that, dissipating considerable amounts of energy in the form of waste heat. Because the efficiency of an amplifier is always 100 percent or less, one can never be made to function as a "perpetual motion" device.The requirement of an external source of power is common to all types of amplifiers, electrical and non-electrical. A common example of a non-electrical amplification system would be power steering in an automobile, amplifying the power of the driver's arms in turning the steering wheel to move the front wheels of the car.

The source of power necessary for the amplification comes from the engine. The active device controlling the driver's "input signal" is a hydraulic valve shuttling fluid power from a pump attached to the engine to a hydraulic piston assisting wheel motion. If the engine stops running, the amplification system fails to amplify the driver's arm power and the car becomes very difficult to turn.

AMPLIFIER GAIN

Because amplifiers have the ability to increase the magnitude of an input signal, it is useful to be able to rate an amplifier's amplifying ability in terms of an output/input ratio. The technical term for an amplifier's output/input magnitude ratio is *gain*. As a ratio of equal units (power out/power in, voltage out/voltage in, or current out/current in), gain is naturally a unitless measurement. Mathematically, gain is symbolized by the capital letter "A".

For example, if an amplifier takes in an AC voltage signal measuring 2 volts RMS and outputs an AC voltage of 30 volts RMS, it has an AC voltage gain of 30 divided by 2, or 15:

$$Av = \frac{V_{ouput}}{V_{input}}$$

$$Av = \frac{30\ V}{2V}$$

$$Av = 15.$$

Correspondingly, if we know the gain of an amplifier and the magnitude of the input signal, we can calculate the magnitude of the output. For example, if an amplifier with an AC current gain of 3.5 is given an AC input signal of 28 mA RMS, the output will be 3.5 times 28 mA, or 98 mA:

$$I_{output} = (A_1)(I_{input})$$

$$I_{output} = (3.5)(28\ \text{mA})$$

$$I_{output} = 98\ \text{mA}.$$

In the last two examples I specifically identified the gains and signal magnitudes in terms of "AC." This was intentional, and illustrates an important concept: electronic amplifiers often respond differently to AC and DC input signals, and may amplify them to different extents. Another way of saying this is that amplifiers often amplify *changes* or *variations* in input signal magnitude (AC) at a different ratio than *steady* input signal magnitudes (DC). The specific reasons for this are too complex to explain at this time, but the fact of the matter is worth mentioning. If gain calculations are to be carried out, it must first be understood what type of signals and gains are being dealt with, AC or DC.

Electrical amplifier gains may be expressed in terms of voltage, current, and/or power, in both AC and DC. A summary of gain definitions is as follows. The triangle-shaped "delta" symbol (Δ) represents *change* in mathematics, so "$\Delta V_{output}/\Delta V_{input}$" means "change in output voltage divided by change in input voltage," or more simply, "AC output voltage divided by AC input voltage":

	DC gains	*AC gains*
Voltage	$A_v = \frac{V_{output}}{V_{input}}$	$A_v = \frac{\Delta V_{output}}{\Delta V_{input}}$
Curront	$A_1 = \frac{l_{output}}{l_{input}}$	$A_1 = \frac{\Delta V_{output}}{\Delta V_{input}}$
Power	$A_p = \frac{P_{output}}{P_{input}}$	$A_p = \frac{(\Delta V_{output})(\Delta l_{output})}{(\Delta l_{input})}$

$$A_p = (A_v)(A_1)$$

Δ = "change in . . ."

If multiple amplifiers are staged, their respective gains form an overall gain equal to the product (multiplication) of the individual gains. If a 1 V signal were applied to the input of the gain of 3 amplifier in Figure below a 3 V signal out of the first amplifier would be further amplified by a gain of 5 at the second stage yielding 15 V at the final output.

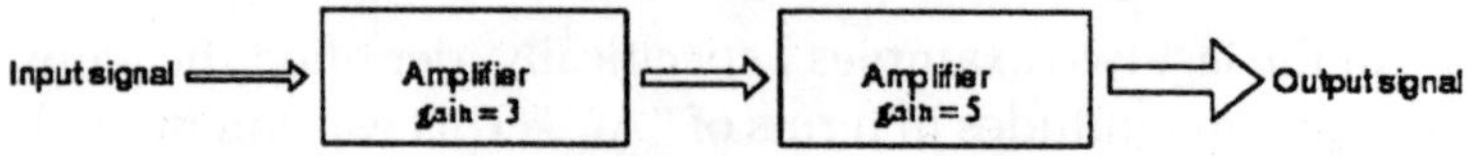

Overall gain = (3)(5) = 15

Fig. The Gain of a Chain of Cascaded Amplifiers is the Product of the Individual Lains.

DECIBELS

In its simplest form, an amplifier's *gain* is a ratio of output over input. Like all ratios, this form of gain is unitless.

However, there is an actual unit intended to represent gain, and it is called the *bel.*

As a unit, the bel was actually devised as a convenient way to represent power *loss* in telephone system wiring rather than *gain* in amplifiers.

The unit's name is derived from Alexander Graham Bell, the famous Scottish inventor whose work was instrumental in developing telephone systems. Originally, the bel represented the amount of signal power loss due to resistance over a standard length of electrical cable. Now, it is defined in terms of the common (base 10) logarithm of a power ratio (output power divided by input power):

$$A_{P(\text{ratio})} = \frac{P_{output}}{P_{input}}$$

$$A_{P(\text{Bel})} = \frac{P_{output}}{P_{input}}.$$

Because the bel is a logarithmic unit, it is nonlinear. To give you an idea of how this works, consider the following table of figures, comparing power losses and gains in bels versus simple ratios:

Table: Gain/loss in bels

Loss/gain as aratio	*Loss/gain inbels*	*Loss/gain as a ratio*	*Loss/gain in bels*
$\frac{P_{output}}{P_{input}}$	$\log \frac{P_{output}}{P_{input}}$	$\frac{P_{output}}{P_{input}}$	$\log \frac{P_{output}}{P_{input}}$
1000	3B	0.1	–1B
100	2B	0.01	–2B
10	1B	0.001	–3B
1 (no loss or gain)	0B	0.0001	–4B

It was later decided that the bel was too large of a unit to be used directly, and so it became customary to apply the metric prefix *deci* (meaning 1/10) to it, making it *deci*bels, or dB. Now, the expression "dB" is so common that many people do not realize it is a combination of "deci-" and "-bel," or that there even is such a unit as the "bel." To put this into perspective, here is another table contrasting power gain/loss ratios against decibels:

Table: Gain/loss in bels

Loss/gain as aratio	*Loss/gain in decibels*	*Loss/gain as a ratio*	*Loss/gain in decibels*
$\frac{P_{output}}{P_{input}}$	$10 \log \frac{P_{output}}{P_{input}}$	$\frac{P_{output}}{P_{input}}$	$10 \log \frac{P_{output}}{P_{input}}$
1000	30dB	0.1	–10dB
100	20dB	0.01	–20dB
10	10dB	0.001	–30dB
1 (no loss or gain)	0 dB	0.0001	–40dB

As a logarithmic unit, this mode of power gain expression covers a wide range of ratios with a minimal span in figures. It is reasonable to ask, "why did anyone feel the need to invent a *logarithmic* unit for electrical signal power loss in a telephone system?" The answer is related to the dynamics of human hearing, the perceptive intensity of which is logarithmic in nature.

Human hearing is highly nonlinear: in order to double the perceived intensity of a sound, the actual sound power must be multiplied by a factor of ten. Relating telephone signal power loss in terms of the logarithmic "bel" scale makes perfect sense in this context: a power loss of 1 bel translates to a perceived sound loss of 50 percent, or 1/2. A power gain of 1 bel translates to a doubling in the perceived intensity of the sound.

An almost perfect analogy to the bel scale is the Richter scale used to describe earthquake intensity: a 6.0 Richter earthquake is 10 times more powerful than a 5.0 Richter earthquake; a 7.0 Richter earthquake 100 times more powerful than a 5.0 Richter earthquake; a 4.0 Richter earthquake is 1/10 as powerful as a 5.0 Richter earthquake, and so on. The measurement scale for chemical pH is likewise logarithmic, a difference of 1 on the scale is equivalent to a tenfold difference in hydrogen ion concentration of a chemical solution. An advantage of using a logarithmic measurement scale is the tremendous range of expression afforded by a relatively small span of numerical values, and it is this advantage which secures the use of Richter numbers for earthquakes and pH for hydrogen ion activity.

Another reason for the adoption of the bel as a unit for gain is for simple expression of system gains and losses. Consider the last system example where two amplifiers were connected tandem to amplify a signal. The respective gain for each amplifier was expressed as a ratio, and the overall gain for the system was the product (multiplication) of those two ratios:

$$\text{Overall gain} = (3)(5) = 15$$

If these figures represented *power* gains, we could directly apply the unit of bels to the task of representing the gain of each amplifier, and of the system altogether.

$$A_{P(Bel)} = \log A_{P(ratio)}$$

$$A_{P(Bel)} = \log 3 \qquad A_{P(Bel)} = \log 5$$

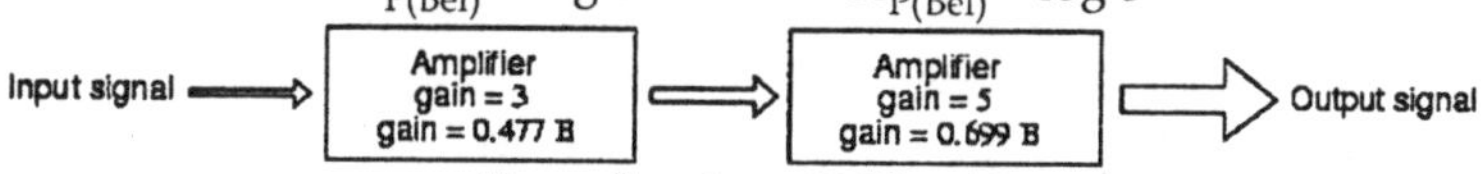

$$\text{Overall gain} = (3)(5) = 15$$

$$\text{Overall gain}_{(Bel)} = \log 15 = 1.176 \text{ B}$$

Fig. Power Gain in Bels is Additive: *0.477 B + 0.699 B = 1.176 B.*

Close inspection of these gain figures in the unit of "bel" yields a discovery: they're additive. Ratio gain figures are multiplicative for staged amplifiers, but gains expressed in bels *add* rather than *multiply* to equal the overall system gain. The first amplifier with its power gain of 0.477 B adds to the second amplifier's power gain of 0.699 B to make a system with an overall power gain of 1.176 B.

Recalculating for decibels rather than bels, we notice the same phenomenon.

$$A_{P(Bel)} = \log A_{P(ratio)}$$

$$A_{P(Bel)} = 10 \log 3 \qquad A_{P(Bel)} = 10 \log 5$$

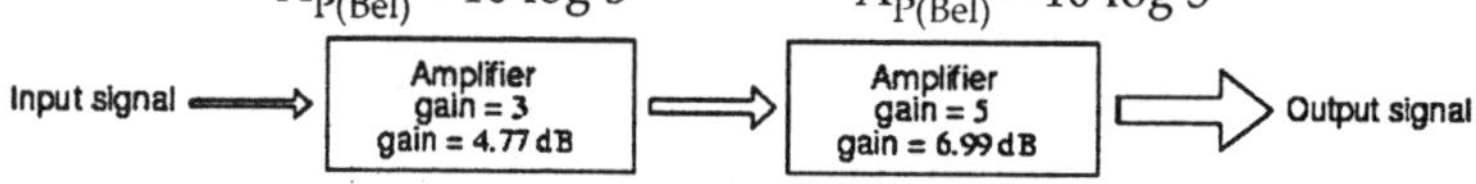

$$\text{Overall gain} = (3)(5) = 15$$

$$\text{Overall gain}_{(dB)} = 10 \log 15 = 11.176 \text{ dB}$$

Fig. Gain of Amplifier Stages in Decibels is Additive: *4.77 dB + 6.99 dB = 11.76 dB.*

To those already familiar with the arithmetic properties of logarithms, this is no surprise. It is an elementary rule of algebra that the antilogarithm of the sum of two numbers' logarithm values equals the product of the two original numbers. In other words, if we take two numbers and determine the logarithm of each, then add those two logarithm figures together, then determine the "antilogarithm" of that sum (elevate the base number of the logarithm — in this case, 10 — to the power of that sum), the result will be the same as if we had simply multiplied the two original numbers together. This algebraic rule forms the heart of a device called a *slide rule,* an analog computer which could, among other things, determine the products and quotients of numbers by addition (adding together physical lengths marked on sliding wood, metal, or plastic scales).

Given a table of logarithm figures, the same mathematical trick could be used to perform otherwise complex multiplications and divisions by only having to do additions and subtractions, respectively. With the advent of high-speed, handheld, digital calculator devices, this elegant calculation technique virtually disappeared from popular use. However, it is still important to understand when working with measurement scales that are logarithmic in nature, such as the bel (decibel) and Richter scales.

When converting a power gain from units of bels or decibels to a unitless ratio, the mathematical inverse function of common logarithms is used: powers of 10, or the *antilog.*

If:

$$A_{P(\text{bel})} = A_{P(\text{ratio})}$$

Then:

$$A_{P(\text{ratio})} = 10\text{A}_{\text{P(Bel)}}$$

Converting decibels into unitless ratios for power gain is much the same, only a division factor of 10 is included in the exponent term:

If:

$$AP(\text{dB}) = 10 \log A_{P(\text{ratio})}$$

Then:

$$A_{P(\text{ratio})} = \frac{A_{P(dB)}}{10^{10}}.$$

Example: Power into an amplifier is 1 watts, the power out is 10 watt. Find the power gain in dB.

$A_{P(dB)} = 10 \log_{10}(P_O/P_I) = 10 \log_{10} (10/1) = 10 \log_{10} (10) = 10 (1) = 10$ dB.

Example: Find the power gain ratio $A_{P(ratio)} = (P_O/P_I)$ for a 20 dB Power gain.

$$A_{P(dB)} = 20 = 10 \log_{10} A_{P(ratio)}$$
$$20/10 = \log_{10} A_{P(ratio)}$$
$$10^{20/10} = 10^{\log_{10}(A_{P(ratio)})}$$
$$100 = A_{P(ratio)} = (P_O/P_I)$$

Because the bel is fundamentally a unit of *power* gain or loss in a system, voltage or current gains and losses don't convert to bels or dB in quite the same way. When using bels or decibels to express a gain other than power, be it voltage or current, we must perform the calculation in terms of how much power gain there would be for that amount of voltage or current gain. For a constant load impedance, a voltage or current gain of 2 equates to a power gain of 4 (2^2); a voltage or current gain of 3 equates to a power gain of 9 (3^2). If we multiply either voltage or current by a given factor, then the power gain incurred by that multiplication will be the square of that factor. This relates back to the forms of Joule's Law where power was calculated from either voltage or current, and resistance:

$$P = \frac{E^2}{R}$$
$$P = 1^2 R$$

Power is proportional to the square of either voltage or current

Thus, when translating a voltage or current gain *ratio* into a respective gain in terms of the bel unit, we must include this exponent in the equation(s):

$$A_{p(Bel)} = \log A_{p(ratio)}$$
$$A_{v(Bel)} = \log A_{v(ratio)}^2 \quad \text{Exponent required}$$
$$A_{I(Bel)} = \log A_{I(ratio)}^2$$

The same exponent requirement holds true when expressing voltage or current gains in terms of decibels:

$$A_{p(dB)} = 10 \log A_{p(ratio)}$$

$$A_{v(dB)} = 10 \log A_{v(ratio)}^2$$ ← Exponent required

$$A_{I(dBI)} = 10 \log A_{I(ratio)}^2$$

However, thanks to another interesting property of logarithms, we can simplify these equations to eliminate the exponent by including the "2" as a *multiplying factor* for the logarithm function. In other words, instead of taking the logarithm of the *square* of the voltage or current gain, we just multiply the voltage or current gain's logarithm figure by 2 and the final result in bels or decibels will be the same:

For bels:

$Av(Bel) = \log A_{V(ratio)^2}$ . . .is the same as . . . $AV(Bel) = 2 \log A_{V(ratio)}$

$A1(Bel) = \log A1_{(ratio)^2}$ . . . is the same as . . . $A1(Bel) = 2 \log A_{1(ratio)}$

For decibels:

$A_{V(dB)} = 10 \log A_{V(ratio)}^2$ . . .is the same as . . . $A_{V(dB)} = 20 \log A_{V(ratio)}$

$A_{1(dB)} = 10 \log A(ratio)^2$ . . . is the same as . . . $A_{1(dB)} = 2 \log A_{1(ratio)}$

The process of converting voltage or current gains from bels or decibels into unitless ratios is much the same as it is for power gains:

If:

$$A_{V(bel)} = 2 \log A_{V(ratio)} \qquad A_{1(Bel)} = 2 \log A_{1(ratio)}$$

Then:

$$A_{V(ratio)} = \frac{A_{V(Bel)}}{10^2} \qquad A_{1(ratio)} = \frac{A_{1(Bel)}}{10^2}$$

Here are the equations used for converting voltage or current gains in decibels into unitless ratios:

If:

$$A_{V(dB)} = 20 \log A_{V(ratio)} \qquad A_{1(dB)} = 20 \log A_{1(ratio)}$$

Then:

$$A_{V(ratio)} = \frac{A_{V(dB)}}{10^{20}} \qquad A_{1(ratio)} = \frac{A_{1(dB)}}{10^{20}}$$

While the bel is a unit naturally scaled for power, another logarithmic unit has been invented to directly express voltage or current gains/losses, and it is based on the *natural* logarithm rather than the *common* logarithm as bels and decibels are. Called the *neper*, its unit symbol is a lower-case "n."

$$A_{V(ratio)} = \frac{V_{output}}{V_{input}} \qquad A_{1(ratio)} = \frac{1_{output}}{1_{input}}$$

$$A_{V(neper)} = \ln A_{V(ratio)} \qquad A_{1(neper)} = \ln A_{1(ratio)}$$

For better or for worse, neither the neper nor its attenuated cousin, the *decineper*, is popularly used as a unit in American engineering applications.

Example: The voltage into a 600 Ω audio line amplifier is 10 mV, the voltage across a 600 Ω load is 1 V. Find the power gain in dB.

$A_{(dB)} = 20 \log_{10}(V_O/V_I) = 20 \log_{10} (1 /0.01) = 20 \log_{10} (100) = 20 (2) = 40$ dB

Example: Find the voltage gain ratio $A_{V(ratio)} = (V_O/V_I)$ for a 20 dB gain amplifier having a 50 Ω input and out impedance.

$$A_{V(dB)} = 20 \log_{10} A_{V(ratio)}$$

$$20 = 20 \log_{10} A_{V(ratio)}$$

$$20/20 = \log_{10} A_{P(ratio)}$$

$$10^{20/20} = 10^{\log_{10} (A_{V(ratio)})}$$

$$10 = A_{V(ratio)} = (V_O/V_I)$$

ABSOLUTE DB SCALES

It is also possible to use the decibel as a unit of absolute power, in addition to using it as an expression of power gain or loss. A common example of this is the use of decibels as a measurement of sound pressure intensity. In cases like these, the measurement is made in reference to some standardized power level defined as 0 dB.

For measurements of sound pressure, 0 dB is loosely defined as the lower threshold of human hearing, objectively quantified as 1 picowatt of sound power per square meter of area.

A sound measuring 40 dB on the decibel sound scale would be 10^4 times greater than the threshold of hearing. A 100 dB sound would be 10^{10} (ten billion) times greater than the threshold of hearing.

Because the human ear is not equally sensitive to all frequencies of sound, variations of the decibel sound-power scale have been developed to represent physiologically equivalent sound intensities at different frequencies. Some sound intensity instruments were equipped with filter networks to give disproportionate indications across the frequency scale, the intent of which to better represent the effects of sound on the human body. Three filtered scales became commonly known as the "A," "B," and "C" weighted scales. Decibel sound intensity indications measured through these respective filtering networks were given in units of dBA, dBB, and dBC. Today, the "A-weighted scale" is most commonly used for expressing the equivalent physiological impact on the human body, and is especially useful for rating dangerously loud noise sources.

Another standard-referenced system of power measurement in the unit of decibels has been established for use in telecommunications systems. This is called the *dBm* scale.

The reference point, 0 dBm, is defined as 1 milliwatt of electrical power dissipated by a 600 Ω load. According to this scale, 10 dBm is equal to 10 times the reference power, or 10 milliwatts; 20 dBm is equal to 100 times the reference power, or 100 milliwatts. Some AC voltmeters come equipped with a dBm range or scale (sometimes labeled "DB") intended for use in measuring AC signal power across a 600 Ω load. 0 dBm on this scale is, of course, elevated above zero because it represents something greater than 0 (actually, it represents 0.7746 volts across a 600 Ω load, voltage being equal to the square root of power times resistance; the square root of 0.001 multiplied by 600). When viewed on the face of an analog meter movement, this dBm scale appears compressed on the left side and expanded on the right in a manner not unlike a resistance scale, owing to its logarithmic nature.

Radio frequency power measurements for low level signals encountered in radio receivers use dBm measurements referenced to a 50 Ω load. Signal generators for the evaluation of radio receivers may output an adjustable dBm rated signal. The signal level is selected by a device called an attenuator, described in the next section.

Table: Absoluts Power Levels in dBm (decibel millwatt)

Power in watts	Power in milliwatts	Power in dBm	Power in milliwatts	Power in dBm
1	1000	30dB	1	0 dB
0.1	100	20dB	0.1	–10dB
0.01	10	10dB	0.01	–20dB
0.004	4	6 dB	0.001	–30dB
0.002	2	3 dB	0.0001	–40dB

Absolute power levels in dBm (decibels referenced to 1 milliwatt: An adaptation of the dBm scale for audio signal strength is used in studio recording and broadcast engineering for standardizing volume levels, and is called the *VU* scale. VU meters are frequently seen on electronic recording instruments to indicate whether or not the recorded signal exceeds the maximum signal level limit of the device, where significant distortion will occur.

When relatively large signals are dealt with, and an absolute dB scale would be useful for representing signal level, specialized decibel scales are sometimes used with reference points greater than the 1 mW used in dBm. Such is the case for the *dBW* scale, with a reference point of 0 dBW established at 1 watt. Another absolute measure of power called the *dBk* scale references 0 dBk at 1 kW, or 1000 watts.

ATTENUATORS

Attenuators are passive devices. It is convenient to discuss them along with decibels. Attenuators weaken or *attenuate* the high level output of a signal generator, for example, to provide a lower level signal for something like the antenna input of a sensitive radio receiver. The attenuator could be built into the signal generator, or be a stand-alone device. It could provide a fixed or adjustable amount of attenuation. An attenuator section can also provide isolation between a source and a troublesome load.

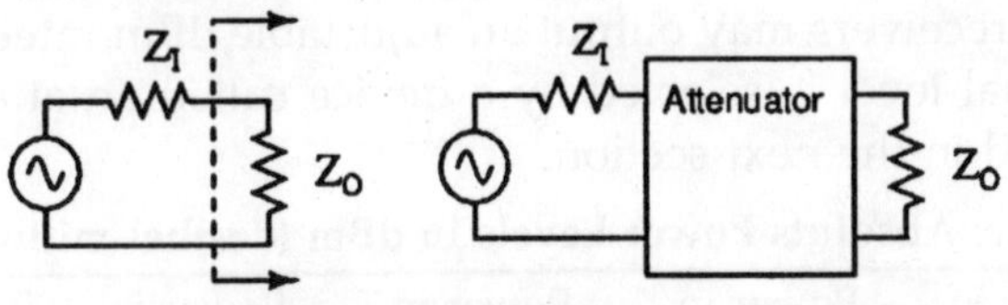

Fig. Constant Impedance Attenuator is Matched to Source Impedance Z_I and Load Impedance Z_O. For Radio Frequency Equipment *Z is 50 W.*

In the case of a stand-alone attenuator, it must be placed in series between the signal source and the load by breaking open the signal path. In addition, it must match both the source impedance Z_I and the load impedance Z_O, while providing a specified amount of attenuation. and load impedances are equal.

Not considered in this section, unequal source and load impedances may be matched by an attenuator section. However, the formulation is more complex.

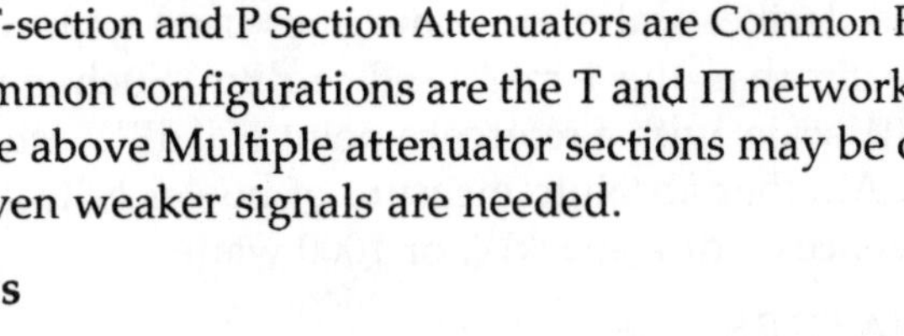

Fig. T-section and P Section Attenuators are Common Forms.

Common configurations are the T and Π networks shown in Figure above Multiple attenuator sections may be cascaded when even weaker signals are needed.

Decibels

Voltage ratios, as used in the design of attenuators are often expressed in terms of decibels. The voltage ratio (K below) must be derived from the attenuation in decibels. Power ratios expressed as decibels are additive. For example, a 10 dB attenuator followed by a 6 dB attenuator provides 16dB of attenuation overall.

$$10 \text{ dB} + 6 \text{ db} = 16 \text{ dB}$$

Changing sound levels are perceptible roughly proportional to the logarithm of the power ratio (P_I/P_O).

$$\text{Sound level} = \log_{10}(P_I/P_O)$$

A change of 1 dB in sound level is barely perceptible to a listener, while 2 db is readily perceptible. An attenuation of 3 dB corresponds to cutting power in half, while a gain of 3 db corresponds to a doubling of the power level. A gain of -3 dB is the same as an attenuation of +3 dB, corresponding to half the original power level.

The power change in decibels in terms of power ratio is:

$$dB = 10 \log_{10}(P_I/P_O)$$

Assuming that the load R_I at P_I is the same as the load resistor R_O at P_O $(R_I = R_O)$, the decibels may be derived from the voltage ratio (V_I/V_O) or current ratio (I_I/I_O):

$$P_O = V_O I_O = V_O^2/R = I_O^2 R$$

$$P_I = V_I I_I = V_I^2/R = I_I^2 R$$

$$dB = 10 \log_{10}(P_I/P_O) = 10 \log_{10}(V_I^2/V_O^2) = 20 \log_{10}(V_I/V_O)$$

$$dB = 0 \log_{10}(P_I/P_O) = 10 \log_{10}(I_I^2/I_O^2) = 20 \log_{10}(I_I/I_O)$$

The two most often used forms of the decibel equation are:

$$dB = 10 \log_{10}(P_I/P_O) \text{ or } dB = 20 \log_{10}(V_I/V_O)$$

We will use the latter form, since we need the voltage ratio. Once again, the voltage ratio form of equation is only applicable where the two corresponding resistors are equal. That is, the source and load resistance need to be equal.

Example: Power into an attenuator is 10 watts, the power out is 1 watt. Find the attenuation in dB.

$$dB = 10 \log_{10}(P_I/P_O) = 10 \log_{10}(10/1) = 10 \log_{10}(10) = 10(1) = 10 \text{ dB}$$

Example: Find the voltage attenuation ratio $(K= (V_I/V_O))$ for a 10 dB attenuator.

$$dB = 10 = 20 \log_{10}(V_I/V_O)$$

$$10/20 = \log_{10}(V_I/V_O)$$

$$10^{10/20} = 10^{\log_{10}(V_I/V_O)}$$

$$3.16 = (V_I/V_O) = A_{P(ratio)}$$

Example: Power into an attenuator is 100 milliwatts, the power out is 1 milliwatt. Find the attenuation in dB.

$dB = 10 \log_{10}(P_I/P_O) = 10 \log_{10}(100/1) = 10 \log_{10}(100) = 10\,(2) = 20$ dB

Example: Find the voltage attenuation ratio ($K = (V_I/V_O)$) for a 20 dB attenuator.

$$dB = 20 = 20 \log_{10}(V_I/V_O)$$

$$10^{20/20} = 10^{\log_{10}(V_I/V_O)}$$

$$10 = (V_I/V_O) = K$$

T-section Attenuator: The T and Π attenuators must be connected to a Z source and Z load impedance. The Z-(arrows) pointing away from the attenuator. The Z-(arrows) pointing toward the attenuator indicates that the impedance seen looking into the attenuator with a load Z on the opposite end is Z, Z=50 Ω for our case. This impedance is a constant (50 Ω) with respect to attenuation– impedance does not change when attenuation is changed.

The table in Figure below lists resistor values for the T and Π attenuators to match a 50 Ω source/ load, as is the usual requirement in radio frequency work.

Telephone utility and other audio work often requires matching to 600 Ω. Multiply all R values by the ratio (600/50) to correct for 600 Ω matching. Multiplying by 75/50 would convert table values to match a 75 Ω source and load.

dB = attenuation in decibels

Z = source/load impedance (resistive)

$K > 1$

$$K = \frac{V_1}{V_0} = 10^{dB/20}$$

$$R_1 = Z\left(\frac{K-1}{K+1}\right)$$

$$R_2 = Z\left(\frac{2K}{K^2-1}\right)$$

Resistors for T–section			
z = 50			
Attenuation			
dB	K=Vi/Vo	R1	R2
1.0	1.12	2.88	433.34
2.0	1.26	5.73	215.24
3.0	1.41	8.55	141.93
4.0	1.58	11.31	104.83
6.0	2.00	16.61	66.93
10.0	3.16	25.97	35.14
20.0	10.00	40.91	10.10

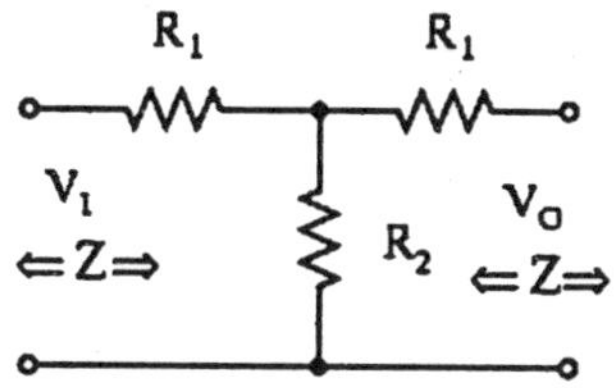

T attenuator

Fig. Formulas for T-section Attenuator Resistors, Given K, the Voltage Attenuation Ratio, and $Z_I = Z_O = 50$ *W*.

The amount of attenuation is customarily specified in dB (decibels). Though, we need the voltage (or current) ratio K to find the resistor values from equations. See the dB/20 term in the power of 10 term for computing the voltage ratio K from dB, above.

The T (and below Π) configurations are most commonly used as they provide bidirectional matching. That is, the attenuator input and output may be swapped end for end and still match the source and load impedances while supplying the same attenuation.

Disconnecting the source and looking in to the right at V_I, we need to see a series parallel combination of R_O, R_I, R_O, and Z looking like an equivalent resistance of Z, the source/ load impedance: (Z is connected to the output.)

$$Z = R_O + (R_2 \,||\, (R_1 + Z))$$

For example, substitute the 10 dB values from the 50 Ω attenuator table for R_1 and R_2 as shown in Figure below.

$$Z = 25.97 + (35.14 \,||\, (25.97 + 50))$$

$$Z = 25.97 + (35.14 \,||\, 75.97)$$

$$Z = 25.97 + 24.03 = 50$$

This shows us that we see 50 Ω looking right into the example attenuator with a 50 Ω load.

Replacing the source generator, disconnecting load Z at V_O, and looking in to the left, should give us the same equation as above for the impedance at V_O, due to symmetry. Moreover, the three resistors must be values which supply the required attenuation from input to output. This is accomplished by the equations for R_1 and R_2 above as applied to the T-attenuator below.

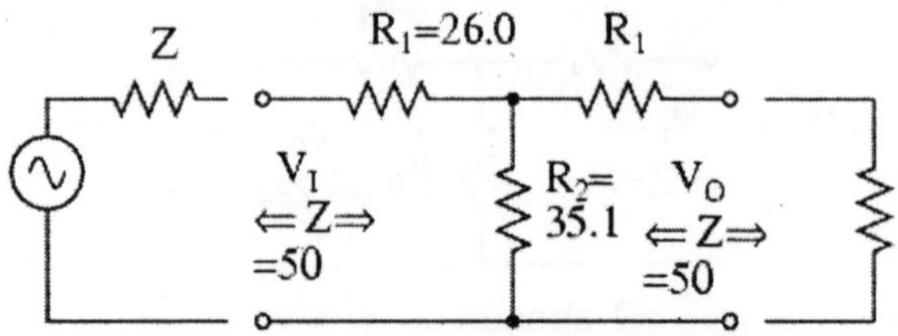

Fig. Tattenuator: 10 dB Attenuators for Matching Input/Output to Z = 50 W

10 dB T-section attenuator for insertion between a 50 Ω source and load.

PI-section Attenuator

The table in Figure below lists resistor values for the Π attenuator matching a 50 Ω source/ load at some common attenuation levels. The resistors corresponding to other attenuation levels may be calculated from the equations.

dB= attenuation in decibels

dB = attenuation in decibels

Z = source/load impedance (resistive)

$K > 1$

$$K = \frac{V_1}{V_0} = 10^{dB/20}$$

$$R_3 = Z\left(\frac{2K}{K^2 - 1}\right)$$

$$R_4 = Z\left(\frac{K-1}{K+1}\right)$$

Resistors for Π–section			
z = 50.00			
Attenuation			
dB	K=Vi/Vo	R3	R4
1.0	1.12	5.77	869.55
2.0	1.26	11.61	436.21
3.0	1.41	17.61	292.40
4.0	1.58	23.85	220.97
6.0	2.00	37.35	150.48
10.0	3.16	7115	96.25
20.0	10.00	247.50	61.11

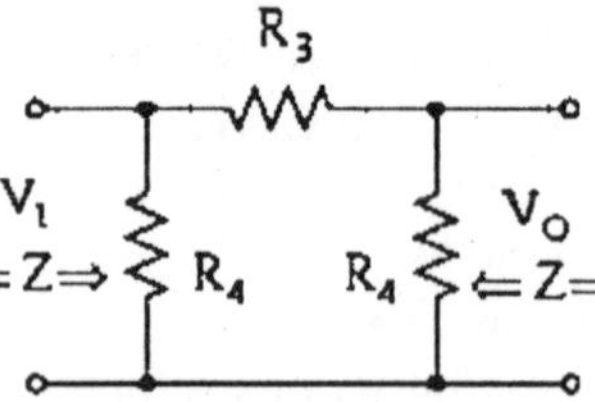

Π attenuator

Fig. Formulas for P-Section Attenuator Resistors, Given K, the Voltage Attenuation Ratio, and $Z_I = Z_O = 50$ W.

The above apply to the Π-attenuator below. What resistor values would be required for both the Π attenuators for 10 dB of attenuation matching a 50 Ω source and load?

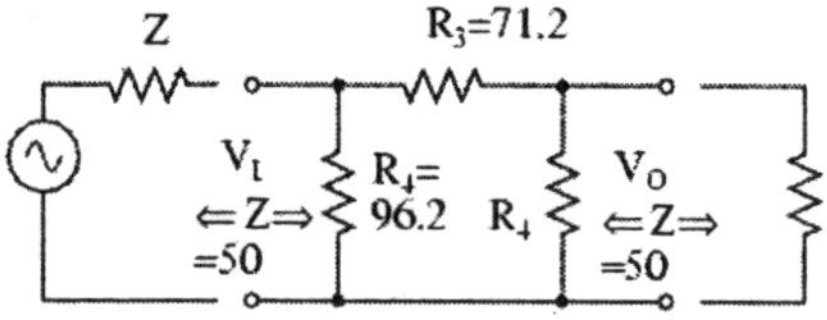

Π attenuator

Fig. 10 dB Attenuator for Matching Input/Output to Z = 50 W. 10 dB P-section attenuator example for matching a 50 W source and load.

The 10 dB corresponds to a voltage attenuation ratio of K=3.16 in the next to last line of the above table. Transfer the resistor values in that line to the resistors.

L-section attenuator

The table in Figure below lists resistor values for the L attenuators to match a 50 Ω source/ load. The table in Figure below lists resistor values for an alternate form. Note that the resistor values are not the same.

dB = attenuation in decibels

Z = source/load impedance (resistive)

$K > 1$

$$K = \frac{V_1}{V_0} = 10^{dB/20}$$

$$R_5 = Z\left(\frac{K-1}{K}\right)$$

$$R_6 = \frac{Z}{(K-1)}$$

Resistors for L-section			
Z = 50.00			
Attenuation L			
dB	K=Vi/Vo	R5	R6
1.0	1.12	5.44	409.77
2.0	1.26	10.28	193.11
3.0	1.41	14.60	121.20
4.0	1.58	18.45	85.49
6.0	2.00	24.94	50.24
10.0	3.16	34.19	23.12
20.R5	10.00	45.00	5.56

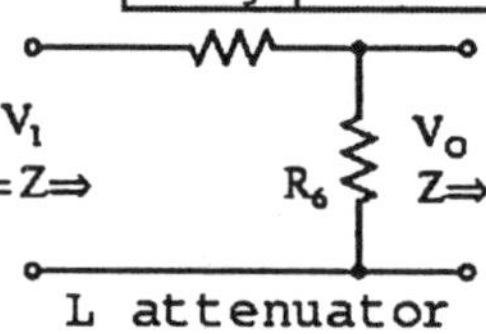

L attenuator

Fig. L-Section Attenuator Table for 50 W Source and Load Impedance.

The above apply to the L attenuator below.

dB = attenuation in decibels

Z = source/load impedance (resistive)

$K > 1$

$$K = \frac{V_1}{V_0} = 10^{dB/20}$$

$$R_7 = Z(K - 1)$$

$$R_8 = Z\frac{K}{(K-1)}$$

Resistors for L–Section			
Z = 50.00			
Attenuation L			
dB	K=Vi/Vo	R7	R8
1.0	1.12	6.10	459.77
2.0	1.26	12.95	243.11
3.0	1.41	20.63	171.20
4.0	1.58	29.24	135.49
6.0	2.00	49.76	100.24
10.0	3.16	108.11	73.12
20.0	10.00	450.00	55.56

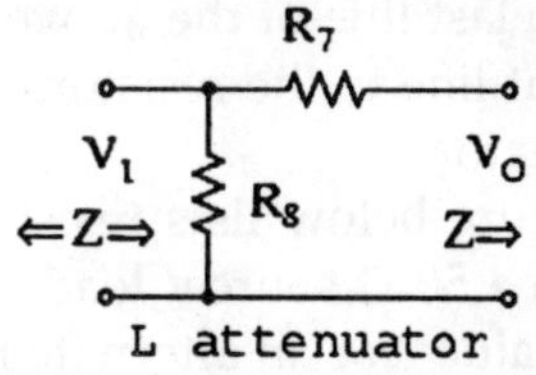

Fig. Alternate form L-Section Attenuator Table for 50 W Source and Load Impedance.

Bridged T Attenuator

The table in Figure below lists resistor values for the bridged T attenuators to match a 50 Ω source and load. The bridged-T attenuator is not often used. Why not?

dB= attenuation in decibels

dB = attenuation in decibels

Z = source/load impedance (resistive)

$K > 1$

$$K = \frac{V_1}{V_0} = 10^{dB/20}$$

$$R_6 = \frac{Z}{(K-1)}$$

$$R_7 = Z(K - 1)$$

Resistors for bridged T			
Z = 50.00			
Attenuation L			
dB	K=Vi/Vo	R7	R6
1.0	1.12	6.10	409.77
2.0	1.26	12.95	193.11
3.0	1.41	20.63	121.20
4.0	1.58	29.24	85.49
6.0	2.00	49.76	50.24
10.0	3.16	108.11	23.12
20.0	10.00	450.00	5.56

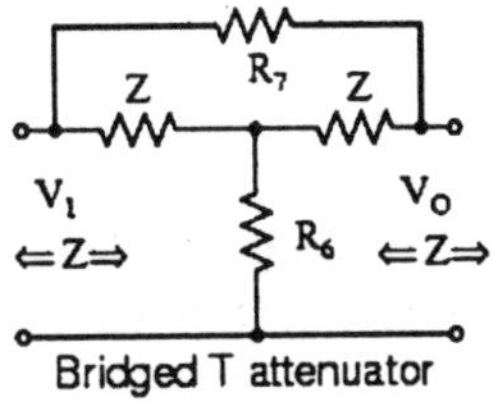

Fig. Formulas and Abbreviated Table for Bridged-T Attenuator Section, Z = *50* W.

Cascaded sections:Attenuator sections can be cascaded as in Figure for more attenuation than may be available from a single section. For example two 10 db attenuators may be cascaded to provide 20 dB of attenuation, the dB values being additive. The voltage attenuation ratio K or V_I/V_O for a 10 dB attenuator section. The voltage attenuation ratio for the two cascaded sections is the product of the two Ks or 3.16x3.16=10 for the two cascaded sections.

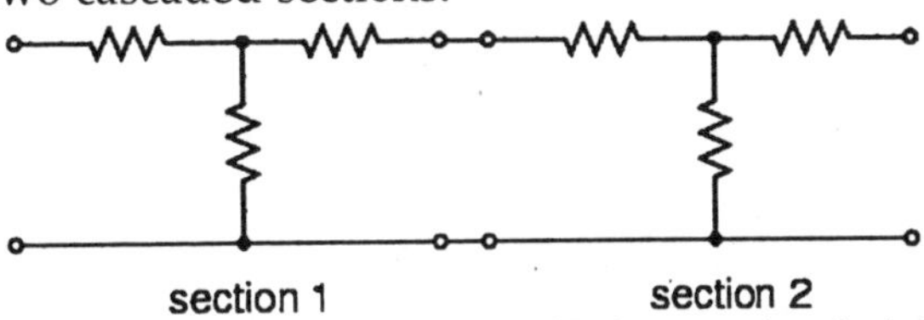

Fig. Cascaded Attenuator Sections: dB Attenuation is Additive.

Variable attenuation can be provided in discrete steps by a switched attenuator. The example Figure below, shown in the 0 dB position, is capable of 0 through 7 dB of attenuation by additive switching of none, one or more sections.

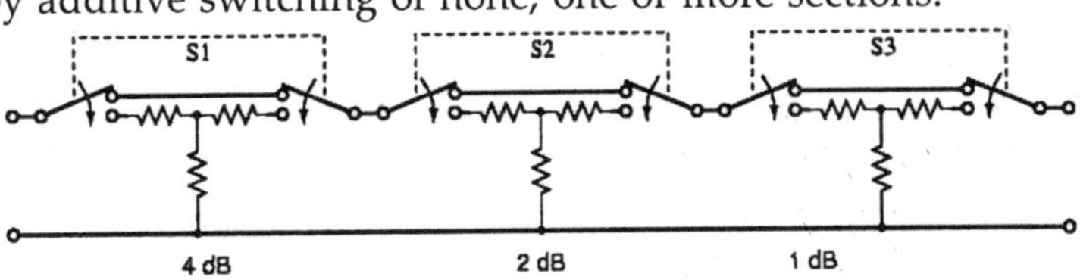

Fig. Switched Attenuator: Attenuation is Variable in Discrete Steps.

The typical multi section attenuator has more sections than the above figure shows. The addition of a 3 or 8 dB section above enables the unit to cover to 10 dB and beyond. Lower signal levels are achieved by the addition of 10 dB and 20 dB sections, or a binary multiple 16 dB section.

RF Attenuators

For radio frequency (RF) work (<1000 Mhz), the individual sections must be mounted in shielded compartments to thwart capacitive coupling if lower signal levels are to be achieved at the highest frequencies. The individual sections of the switched attenuators in the previous section are mounted in shielded sections. Additional measures may be taken to extend the frequency range to beyond 1000 Mhz. This involves construction from special shaped lead-less resistive elements.

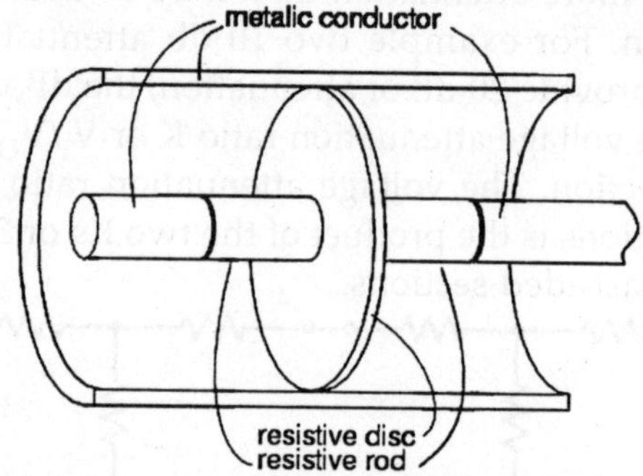

Fig. Coaxial T-Attenuator for Radio Frequency Work.

A coaxial T-section attenuator consisting of resistive rods and a resistive disk is shown in Figure above. This construction is usable to a few gigahertz. The coaxial Π version would have one resistive rod between two resistive disks in the coaxial line as in Figure below.

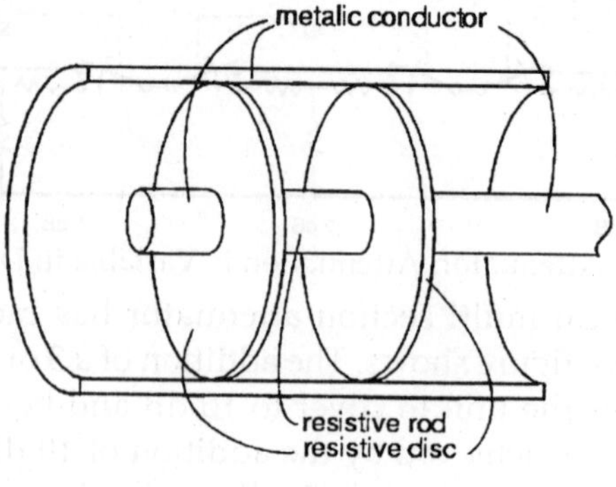

Fig. Coaxial P-Attenuator for Radio Frequency Work.

RF connectors, not shown, are attached to the ends of the above T and Π attenuators. The connectors allow individual attenuators to be cascaded, in addition to connecting between a source and load. For example, a 10 dB attenuator may be placed between a troublesome signal source and an expensive spectrum analyzer input. Even though we may not need the attenuation, the expensive test equipment is protected from the source by attenuating any overvoltage.

Chapter 2

Diodes and Rectifiers

A *diode* is an electrical device allowing current to move through it in one direction with far greater ease than in the other. The most common kind of diode in modern circuit design is the *semiconductor* diode, although other diode technologies exist. Semiconductor diodes are symbolized in schematic diagrams such as Figure below. The term "diode" is customarily reserved for small signal devices, I d" 1 A. The term *rectifier* is used for power devices, I > 1 A.

Semiconductor diode schematic symbol: Arrows indicate the direction of electron current flow.

When placed in a simple battery-lamp circuit, the diode will either allow or prevent current through the lamp, depending on the polarity of the applied voltage.

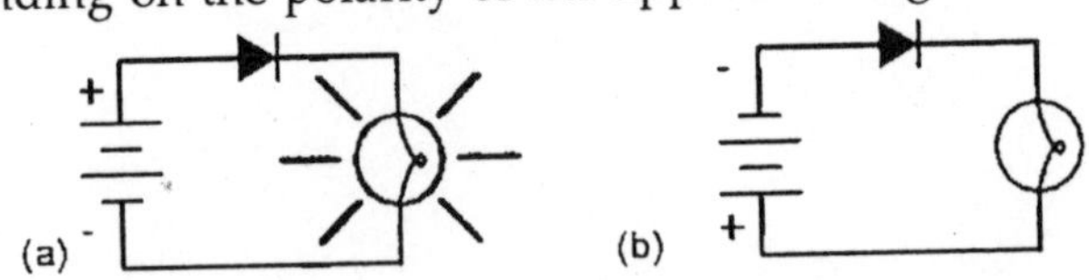

Fig. Diode operation: (a) Current flow is permitted; the diode is forward biased. (b) Current flow is prohibited; the diode is reversed biased.

When the polarity of the battery is such that electrons are allowed to flow through the diode, the diode is said to be *forward-biased*. Conversely, when the battery is "backward" and the diode blocks current, the diode is said to be *reverse-biased*. A diode may be thought of as like a switch: "closed" when forward-biased and "open" when reverse-biased.

Oddly enough, the direction of the diode symbol's "arrowhead" points *against* the direction of electron flow. This is because the diode symbol was invented by engineers, who predominantly use *conventional flow* notation in their schematics, showing current as a flow of charge from the positive (+) side of the voltage source to the negative (-). This convention holds true for all semiconductor symbols possessing "arrowheads:" the arrow points in the permitted direction of conventional flow, and against the permitted direction of electron flow.

Diode behavior is analogous to the behavior of a hydraulic device called a *check valve.* A check valve allows fluid flow through it in only one direction as in Figure below.

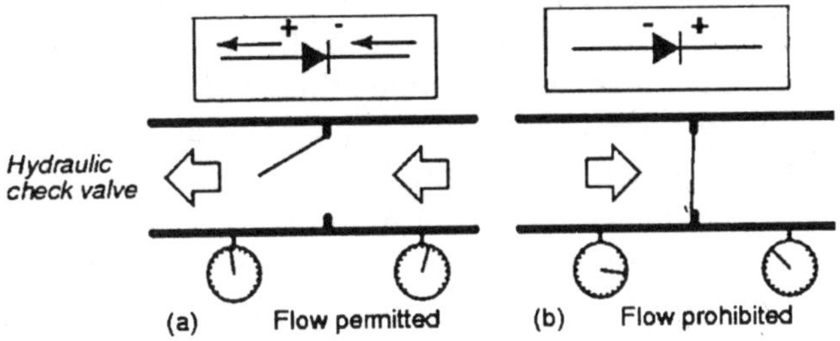

Fig. Hydraulic check valve analogy: (a) Electron current flow permitted. (b) Current flow prohibited.

Check valves are essentially pressure-operated devices: they open and allow flow if the pressure across them is of the correct "polarity" to open the gate (in the analogy shown, greater fluid pressure on the right than on the left). If the pressure is of the opposite "polarity," the pressure difference across the check valve will close and hold the gate so that no flow occurs.

Like check valves, diodes are essentially "pressure-" operated (voltage-operated) devices.

The essential difference between forward-bias and reverse-bias is the polarity of the voltage dropped across the diode.

Let's take a closer look at the simple battery-diode-lamp circuit shown earlier, this time investigating voltage drops across the various components in Figure below.

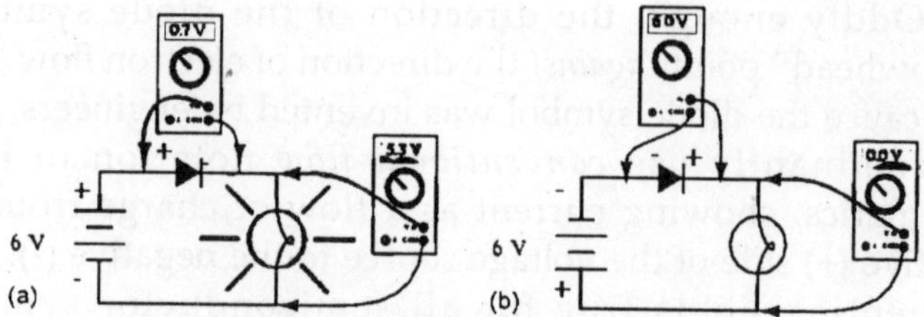

Fig. Diode Circuit Voltage Measurements:
(a) Forward Biased. (b) Reverse Biased.

A forward-biased diode conducts current and drops a small voltage across it, leaving most of the battery voltage dropped across the lamp. If the battery's polarity is reversed, the diode becomes reverse-biased, and drops *all* of the battery's voltage leaving none for the lamp. If we consider the diode to be a self-actuating switch (closed in the forward-bias mode and open in the reverse-bias mode), this behavior makes sense. The most substantial difference is that the diode drops a lot more voltage when conducting than the average mechanical switch (0.7 volts versus tens of millivolts).

This forward-bias voltage drop exhibited by the diode is due to the action of the depletion region formed by the P-N junction under the influence of an applied voltage. If no voltage applied is across a semiconductor diode, a thin depletion region exists around the region of the P-N junction, preventing current flow. The depletion region is almost devoid of available charge carriers, and acts as an insulator:

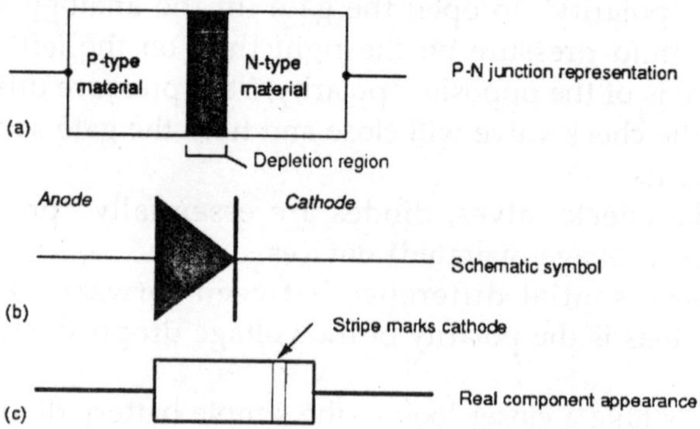

Fig. Diode Representations: PN-junction Model, Schematic Symbol, physical part.

The schematic symbol of the diode is shown in Figure above (b) such that the anode (pointing end) corresponds to the P-type semiconductor at (a). The cathode bar, non-pointing end, at (b) corresponds to the N-type material at (a). Also note that the cathode stripe on the physical part (c) corresponds to the cathode on the symbol.

If a reverse-biasing voltage is applied across the P-N junction, this depletion region expands, further resisting any current through it.

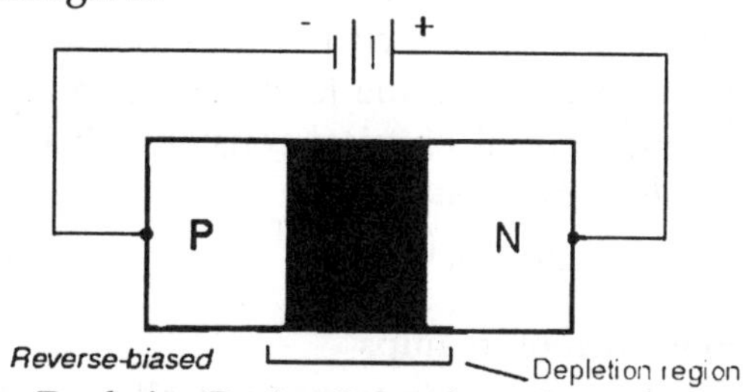

Fig. Depletion Region Expands with Reverse Bias.

Conversely, if a forward-biasing voltage is applied across the P-N junction, the depletion region collapses becoming thinner. The diode becomes less resistive to current through it. In order for a sustained current to go through the diode; though, the depletion region must be fully collapsed by the applied voltage. This takes a certain minimum voltage to accomplish, called the *forward voltage*.

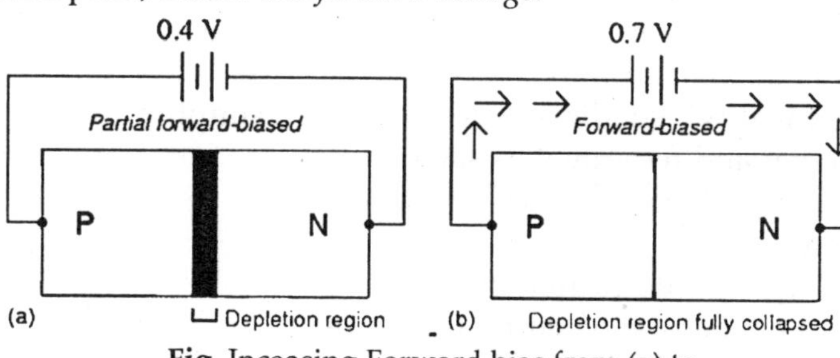

Fig. Inceasing Forward bias from (a) to (b) Decreases Ddepletion Region Thickness.

For silicon diodes, the typical forward voltage is 0.7 volts, nominal. For germanium diodes, the forward voltage is only 0.3 volts. The chemical constituency of the P-N junction comprising the diode accounts for its nominal forward voltage figure, which is why silicon and germanium diodes have such different forward voltages. Forward voltage drop remains approximately constant for a wide range of diode currents, meaning that diode voltage drop is not like that of a resistor or even a normal (closed) switch. For most simplified circuit analysis, the voltage drop across a conducting diode may be considered constant at the nominal figure and not related to the amount of current.

Actually, forward voltage drop is more complex. An equation describes the exact current through a diode, given the voltage dropped across the junction, the temperature of the junction, and several physical constants. It is commonly known as the *diode equation*:

$$I_D = I_S\,(e^{qVD/NkT} - 1)$$

Where,

ID = Diode current in amps

IS = Saturation current in amps (typically 1 × 10–12 amsp)

e = Euler's constant (~2.718281828)

q = charge of electron (1.6 × 10^{-19} coulombs)

V_D = Voltage applied across diode in volts.

k = Boltzmann's constant (1.38 × 10^{-23})

T = Junction temperature in Kelvins

The term kT/q describes the voltage produced within the P-N junction due to the action of temperature, and is called the *thermal voltage*, or V_t of the junction. At room temperature, this is about 26 millivolts. Knowing this, and assuming a "nonideality" coefficient of 1, we may simplify the diode equation and re-write it as such:

$$I_D = IS\,(e^{VD/0.026} - 1)$$

Where,

I_D = Diode current in amps

I_S = Saturation current in amps (typically 1 × 10^{-12} amps)

e = Euler's constant (~ 2.71828128)

V_D = Voltage applied a cross diode in volts.

You need not be familiar with the "diode equation" to analyze simple diode circuits. Just understand that the voltage dropped across a current-conducting diode *does* change with the amount of current going through it, but that this change is fairly small over a wide range of currents. The voltage drop across a conducting, semiconductor diode remains constant at 0.7 volts for silicon and 0.3 volts for germanium. However, some circuits intentionally make use of the P-N junction's inherent exponential current/voltage relationship and thus can only be understood in the context of this equation. Also, since temperature is a factor in the diode equation, a forward-biased P-N junction may also be used as a temperature-sensing device, and thus can only be understood if one has a conceptual grasp on this mathematical relationship.

A reverse-biased diode prevents current from going through it, due to the expanded depletion region. In actuality, a very small amount of current can and does go through a reverse-biased diode, called the *leakage current,* but it can be ignored for most purposes. The ability of a diode to withstand reverse-bias voltages is limited, as it is for any insulator. If the applied reverse-bias voltage becomes too great, the diode will experience a condition known as *breakdown,* which is usually destructive. A diode's maximum reverse-bias voltage rating is known as the *Peak Inverse Voltage,* or *PIV,* and may be obtained from the manufacturer. Like forward voltage, the PIV rating of a diode varies with temperature, except that PIV *increases* with increased temperature and *decreases* as the diode becomes cooler — exactly opposite that of forward voltage.

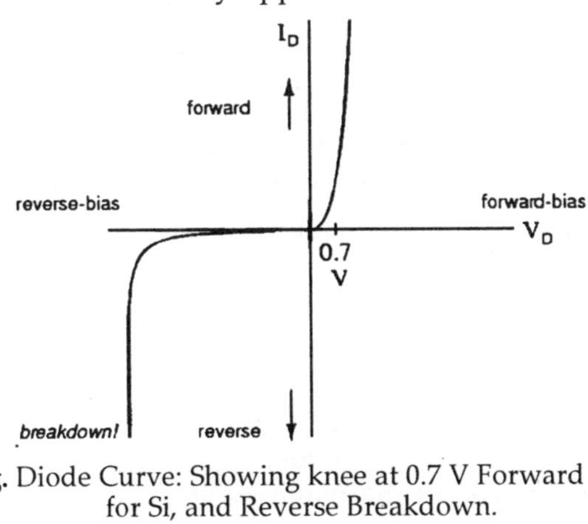

Fig. Diode Curve: Showing knee at 0.7 V Forward Bias for Si, and Reverse Breakdown.

Typically, the PIV rating of a generic "rectifier" diode is at least 50 volts at room temperature. Diodes with PIV ratings in the many thousands of volts are available for modest prices.

DIODE RATINGS

In addition to forward voltage drop (V_f) and peak inverse voltage (PIV), there are many other ratings of diodes important to circuit design and component selection. Semiconductor manufacturers provide detailed specifications on their products — diodes included — in publications known as *datasheets*.

A typical diode datasheet will contain figures for the following parameters:

Maximum repetitive reverse voltage = V_{RRM}, the maximum amount of voltage the diode can withstand in reverse-bias mode, in repeated pulses. Ideally, this figure would be infinite.

Maximum DC reverse voltage = V_R or V_{DC}, the maximum amount of voltage the diode can withstand in reverse-bias mode on a continual basis. Ideally, this figure would be infinite.

Maximum forward voltage = V_F, usually specified at the diode's rated forward current. Ideally, this figure would be zero: the diode providing no opposition whatsoever to forward current. In reality, the forward voltage is described by the "diode equation."

Maximum (average) forward current = $I_{F(AV)}$, the maximum average amount of current the diode is able to conduct in forward bias mode. This is fundamentally a thermal limitation: how much heat can the PN junction handle, given that dissipation power is equal to current (I) multiplied by voltage (V or E) and forward voltage is dependent upon both current and junction temperature. Ideally, this figure would be infinite.

Maximum (peak or surge) forward current = I_{FSM} or $i_{f(surge)}$, the maximum peak amount of current the diode is able to conduct in forward bias mode. Again, this rating is limited by the diode junction's thermal capacity, and is usually much higher than the average current rating due to thermal inertia (the fact that it takes a finite amount of time for the diode to reach maximum temperature for a given current). Ideally, this figure would be infinite.

Maximum total dissipation = P_D, the amount of power (in watts) allowable for the diode to dissipate, given the dissipation (P=IE) of diode current multiplied by diode voltage drop, and also the dissipation ($P=I^2R$) of diode current squared multiplied by bulk resistance. Fundamentally limited by the diode's thermal capacity (ability to tolerate high temperatures).

Operating junction temperature = T_J, the maximum allowable temperature for the diode's PN junction, usually given in degrees Celsius (°C). Heat is the "Achilles' heel" of semiconductor devices: they *must* be kept cool to function properly and give long service life.

Storage temperature range = T_{STG}, the range of allowable temperatures for storing a diode (unpowered). Sometimes given in conjunction with operating junction temperature (T_J), because the maximum storage temperature and the maximum operating temperature ratings are often identical. If anything, though, maximum storage temperature rating will be greater than the maximum operating temperature rating.

Thermal resistance = R(È), the temperature difference between junction and outside air ($R(È)_{JA}$) or between junction and leads ($R(È)_{JL}$) for a given power dissipation. Expressed in units of degrees Celsius per watt (°C/W). Ideally, this figure would be zero, meaning that the diode package was a perfect thermal conductor and radiator, able to transfer all heat energy from the junction to the outside air (or to the leads) with no difference in temperature across the thickness of the diode package.

A high thermal resistance means that the diode will build up excessive temperature at the junction (where its critical) despite best efforts at cooling the outside of the diode, and thus will limit its maximum power dissipation.

Maximum reverse current = I_R, the amount of current through the diode in *reverse-bias* operation, with the maximum rated inverse voltage applied (V_{DC}). Sometimes referred to as *leakage current*. Ideally, this figure would be zero, as a perfect diode would block all current when reverse-biased. In reality, it is very small compared to the maximum forward current.

Typical junction capacitance = C_J, the typical amount of capacitance intrinsic to the junction, due to the depletion region acting as a dielectric separating the anode and cathode connections. This is usually a very small figure, measured in the range of picofarads (pF).

Reverse recovery time = t_{rr}, the amount of time it takes for a diode to "turn off" when the voltage across it alternates from forward-bias to reverse-bias polarity. Ideally, this figure would be zero: the diode halting conduction *immediately* upon polarity reversal. For a typical rectifier diode, reverse recovery time is in the range of tens of microseconds; for a "fast switching" diode, it may only be a few nanoseconds.

Most of these parameters vary with temperature or other operating conditions, and so a single figure fails to fully describe any given rating. Therefore, manufacturers provide graphs of component ratings plotted against other variables (such as temperature), so that the circuit designer has a better idea of what the device is capable of.

RECTIFIER CIRCUITS

Now we come to the most popular application of the diode: *rectification*. Simply defined, rectification is the conversion of alternating current (AC) to direct current (DC). This involves a device that only allows one-way flow of electrons. As we have seen, this is exactly what a semiconductor diode does. The simplest kind of rectifier circuit is the *half-wave* rectifier. It only allows one half of an AC waveform to pass through to the load.

Half-wave rectifier circuit: For most power applications, half-wave rectification is insufficient for the task. The harmonic content of the rectifier's output waveform is very large and consequently difficult to filter. Furthermore, the AC power source only supplies power to the load once every half-cycle, meaning that much of its capacity is unused. Half-wave rectification is, however, a very simple way to reduce power to a resistive load. Some two-position lamp dimmer switches apply full AC power to the lamp filament for "full" brightness and then half-wave rectify it for a lesser light output.

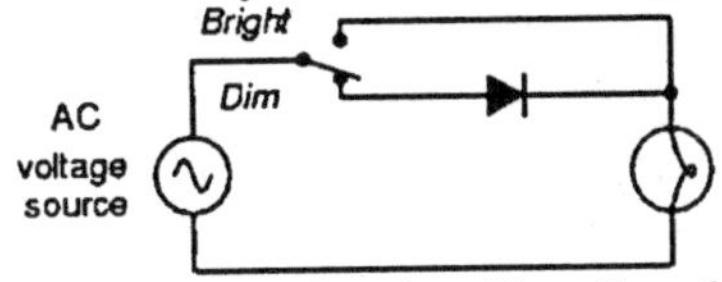

Fig. Half-wave Rectifier Application: Two Level Lamp Dimmer.

In the "Dim" switch position, the incandescent lamp receives approximately one-half the power it would normally receive operating on full-wave AC. Because the half-wave rectified power pulses far more rapidly than the filament has time to heat up and cool down, the lamp does not blink. Instead, its filament merely operates at a lesser temperature than normal, providing less light output. This principle of "pulsing" power rapidly to a slow-responding load device to control the electrical power sent to it is common in the world of industrial electronics.

Since the controlling device (the diode, in this case) is either fully conducting or fully nonconducting at any given time, it dissipates little heat energy while controlling load power, making this method of power control very energy-efficient.

This circuit is perhaps the crudest possible method of pulsing power to a load, but it suffices as a proof-of-concept application.

If we need to rectify AC power to obtain the full use of *both* half-cycles of the sine wave, a different rectifier circuit configuration must be used. Such a circuit is called a *full-wave* rectifier. One kind of full-wave rectifier, called the *center-tap* design, uses a transformer with a center-tapped secondary winding and two diodes.

This circuit's operation is easily understood one half-cycle at a time. Consider the first half-cycle, when the source voltage polarity is positive (+) on top and negative (-) on bottom. At this time, only the top diode is conducting; the bottom diode is blocking current, and the load "sees" the first half of the sine wave, positive on top and negative on bottom. Only the top half of the transformer's secondary winding carries current during this half-cycle as in Figure.

During the next half-cycle, the AC polarity reverses. Now, the other diode and the other half of the transformer's secondary winding carry current while the portions of the circuit formerly carrying current during the last half-cycle sit idle. The load still "sees" half of a sine wave, of the same polarity as before: positive on top and negative on bottom.

One disadvantage of this full-wave rectifier design is the necessity of a transformer with a center-tapped secondary winding. If the circuit in question is one of high power, the size and expense of a suitable transformer is significant. Consequently, the center-tap rectifier design is only seen in low-power applications.

The full-wave center-tapped rectifier polarity at the load may be reversed by changing the direction of the diodes. Furthermore, the reversed diodes can be paralleled with an existing positive-output rectifier. The result is dual-polarity full-wave center-tapped rectifier in Figure. Note that the connectivity of the diodes themselves is the same configuration as a bridge.

Another, more popular full-wave rectifier design exists, and it is built around a four-diode bridge configuration. For obvious reasons, this design is called a *full-wave bridge*.

Current directions for the full-wave bridge rectifier circuit are as shown in Figure below for positive half-cycle and Figure below for negative half-cycles of the AC source waveform. Note that regardless of the polarity of the input, the current flows in the same direction through the load. That is, the negative half-cycle of source is a positive half-cycle at the load. The current flow is through two diodes in series for both polarities. Thus, two diode drops of the source voltage are lost (0.7·2=1.4 V for Si) in the diodes. This is a disadvantage compared with a full-wave center-tap design. This disadvantage is only a problem in very low voltage power supplies.

Remembering the proper layout of diodes in a full-wave bridge rectifier circuit can often be frustrating to the new student of electronics. I've found that an alternative representation of this circuit is easier both to remember and to comprehend. It's the exact same circuit, except all diodes are drawn in a horizontal attitude, all "pointing" the same direction.

One advantage of remembering this layout for a bridge rectifier circuit is that it expands easily into a polyphase version in Figure below.

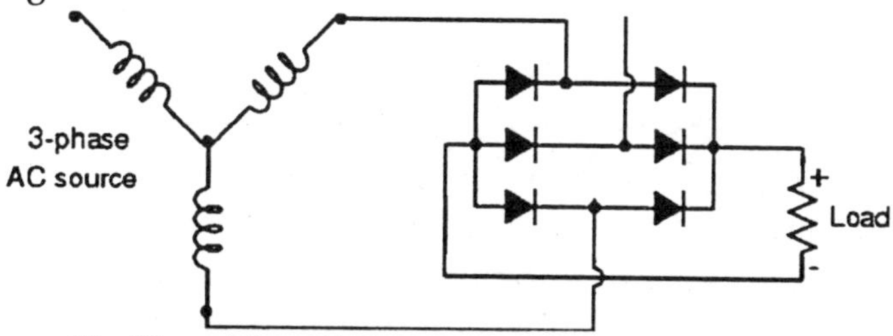

Fig: Three-Phase Full-wave Bridge Rectifier Circuit.

Each three-phase line connects between a pair of diodes: one to route power to the positive (+) side of the load, and the other to route power to the negative (-) side of the load. Polyphase systems with more than three phases are easily accommodated into a bridge rectifier scheme. Take for instance the six-phase bridge rectifier circuit in Figure below.

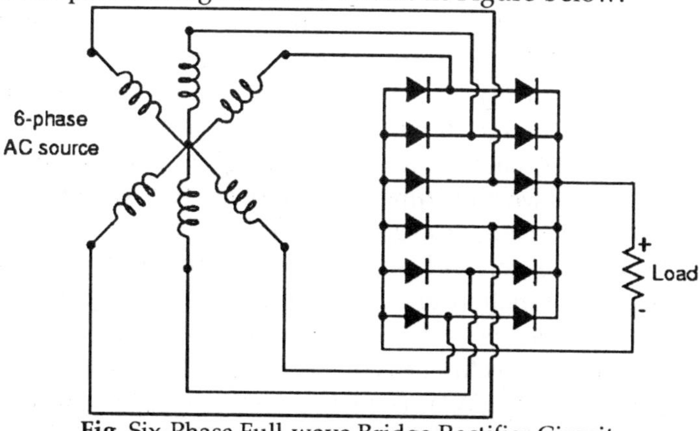

Fig. Six-Phase Full-wave Bridge Rectifier Circuit.

When polyphase AC is rectified, the phase-shifted pulses overlap each other to produce a DC output that is much "smoother" (has less AC content) than that produced by the rectification of single-phase AC. This is a decided advantage in high-power rectifier circuits, where the sheer physical size of filtering components would be prohibitive but low-noise DC power must be obtained. The diagram in Figure below shows the full-wave rectification of three-phase AC.

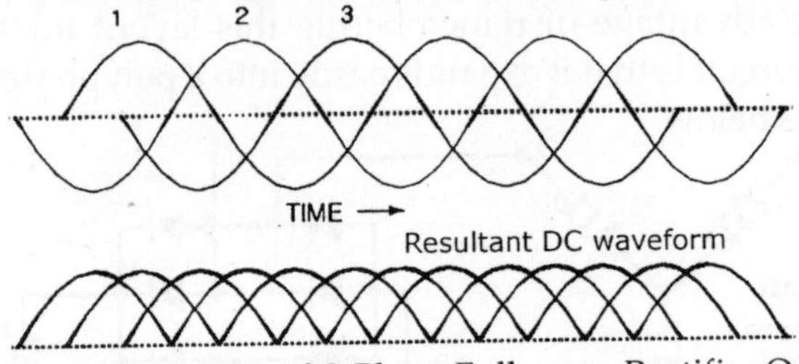

Fig. Three-Phase AC and 3-Phase Full-wave Rectifier Output.

In any case of rectification – single-phase or polyphase – the amount of AC voltage mixed with the rectifier's DC output is called *ripple voltage*. In most cases, since "pure" DC is the desired goal, ripple voltage is undesirable. If the power levels are not too great, filtering networks may be employed to reduce the amount of ripple in the output voltage.

Sometimes, the method of rectification is referred to by counting the number of DC "pulses" output for every 360° of electrical "rotation." A single-phase, half-wave rectifier circuit, then, would be called a *1-pulse* rectifier, because it produces a single pulse during the time of one complete cycle (360°) of the AC waveform. A single-phase, full-wave rectifier (regardless of design, center-tap or bridge) would be called a *2-pulse* rectifier, because it outputs two pulses of DC during one AC cycle's worth of time. A three-phase full-wave rectifier would be called a *6-pulse* unit.

Modern electrical engineering convention further describes the function of a rectifier circuit by using a three-field notation of *phases, ways*, and number of *pulses*. A single-phase, half-wave rectifier circuit is given the somewhat cryptic designation of 1Ph1W1P (1 phase, 1 way, 1 pulse), meaning that the AC supply voltage is single-phase, that current on each phase of the AC supply lines moves in only one direction (way), and that there is a single pulse of DC produced for every 360° of electrical rotation. A single-phase, full-wave, center-tap rectifier circuit would be designated as 1Ph1W2P in this notational system: 1 phase, 1 way or direction of current in each winding half, and 2 pulses or output voltage per cycle. A single-phase, full-wave, bridge rectifier would be designated as 1Ph2W2P: the same as for the center-tap design, except current can go *both* ways through the AC lines instead of just one way. The three-phase bridge rectifier circuit shown earlier would be called a 3Ph2W6P rectifier.

Is it possible to obtain more pulses than twice the number of phases in a rectifier circuit? The answer to this question is yes: especially in polyphase circuits. Through the creative use of transformers, sets of full-wave rectifiers may be paralleled in such a way that more than six pulses of DC are produced for three phases of AC. A 30° phase shift is introduced from primary to secondary of a three-phase transformer when the winding configurations are not of the same type. In other words, a transformer connected either Y-Δ or Δ-Y will exhibit this 30° phase shift, while a transformer connected Y-Y or Δ-Δ will not. This phenomenon may be exploited by having one transformer connected Y-Y feed a bridge rectifier, and have another transformer connected Y-Δ feed a second bridge rectifier, then parallel the DC outputs of both rectifiers. Since the ripple voltage waveforms of the two rectifiers' outputs are phase-shifted 30° from one another, their superposition results in less ripple than either rectifier output considered separately: 12 pulses per 360° instead of just six:

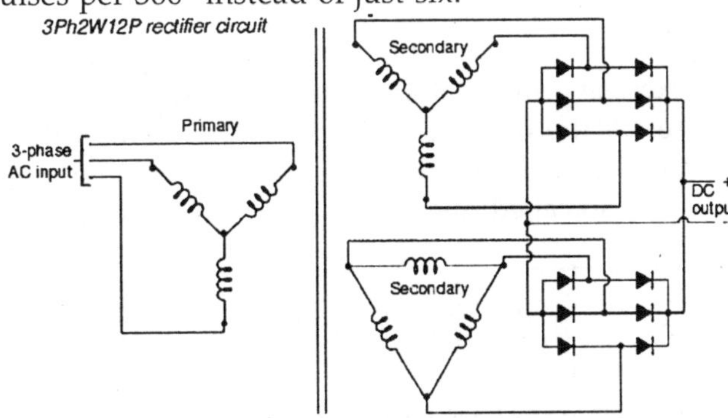

Fig. Polyphase Rectifier Circuit: 3-Phase 2-Way 12-Pulse (3Ph2W12P)

INDUCTOR COMMUTATING CIRCUITS

A popular use of diodes is for the mitigation of inductive "kickback:" the pulses of high voltage produced when direct current through an inductor is interrupted. Take, for example, this simple circuit in Figure with no protection against inductive kickback.

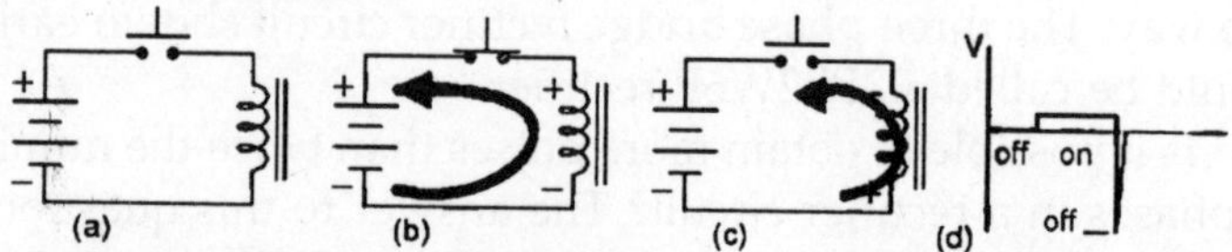

Fig. Inductive kickback: (a) Switch open. (b) Switch closed, electron current flows from battery through coil which has polarity matching battery. Magnetic field stores energy. (c) Switch open, Current still flows in coil due to collapsing magnetic field. Note polarity change on coil changed. (d) Coil voltage vs time.

When the pushbutton switch is actuated, current goes through the inductor, producing a magnetic field around it. When the switch is de-actuated, its contacts open, interrupting current through the inductor, and causing the magnetic field to rapidly collapse. Because the voltage induced in a coil of wire is directly proportional to the *rate of change* over time of magnetic flux (Faraday's Law: e = NdÖ/dt), this rapid collapse of magnetism around the coil produces a high voltage "spike".

If the inductor in question is an electromagnet coil, such as in a solenoid or relay (constructed for the purpose of creating a physical force via its magnetic field when energized), the effect of inductive "kickback" serves no useful purpose at all. In fact, it is quite detrimental to the switch, as it causes excessive arcing at the contacts, greatly reducing their service life. Of the practical methods for mitigating the high voltage transient created when the switch is opened, none so simple as the so-called *commutating diode* in Figure.

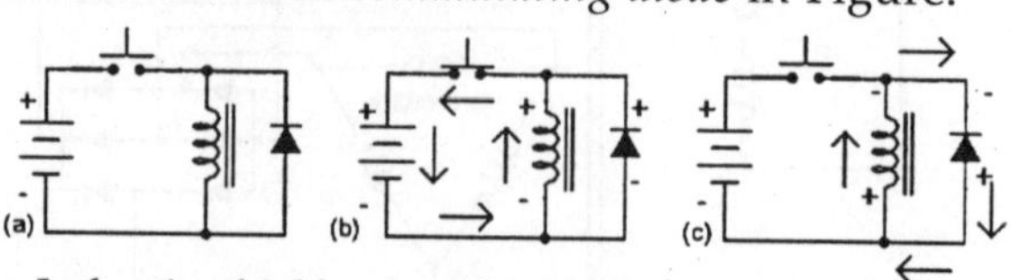

Fig. Inductive kickback with protection: (a) Switch open. (b)Switch closed, storing energy in magnetic field. (c) Switch open, inductive kickback is shorted by diode.

In this circuit, the diode is placed in parallel with the coil, such that it will be reverse-biased when DC voltage is applied to the coil through the switch.

However, when the switch is opened, the coil's inductance responds to the decrease in current by inducing a voltage of reverse polarity, in an effort to maintain current at the same magnitude and in the same direction. This sudden reversal of voltage polarity across the coil forward-biases the diode, and the diode provides a current path for the inductor's current, so that its stored energy is dissipated slowly rather than suddenly.

As a result, the voltage induced in the coil by its collapsing magnetic field is quite low: merely the forward voltage drop of the diode, rather than hundreds of volts as before. Thus, the switch contacts experience a voltage drop equal to the battery voltage plus about 0.7 volts (if the diode is silicon) during this discharge time.

In electronics parlance, *commutation* refers to the reversal of voltage polarity or current direction. Thus, the purpose of a *commutating diode* is to act whenever voltage reverses polarity, for example, on an inductor coil when current through it is interrupted. A less formal term for a commutating diode is *snubber*, because it "snubs" or "squelches" the inductive kickback.

A noteworthy disadvantage of this method is the extra time it imparts to the coil's discharge. Because the induced voltage is clamped to a very low value, its rate of magnetic flux change over time is comparatively slow. Remember that Faraday's Law describes the magnetic flux rate-of-change (dÖ/dt) as being proportional to the induced, instantaneous voltage (*e* or *v*). If the instantaneous voltage is limited to some low figure, then the rate of change of magnetic flux over time will likewise be limited to a low (slow) figure.

If an electromagnet coil is "snubbed" with a commutating diode, the magnetic field will dissipate at a relatively slow rate compared to the original scenario (no diode) where the field disappeared almost instantly upon switch release. The amount of time in question will most likely be less than one second, but it will be measurably slower than without a commutating diode in place. This may be an intolerable consequence if the coil is used to actuate an electromechanical relay, because the relay will possess a natural "time delay" upon coil de-energization, and an unwanted delay of even a fraction of a second may wreak havoc in some circuits.

Unfortunately, one cannot eliminate the high-voltage transient of inductive kickback *and* maintain fast de-magnetization of the coil: Faraday's Law will not be violated. However, if slow de-magnetization is unacceptable, a compromise may be struck between transient voltage and time by allowing the coil's voltage to rise to some higher level (but not so high as without a commutating diode in place). The schematic in Figure shows how this can be done.

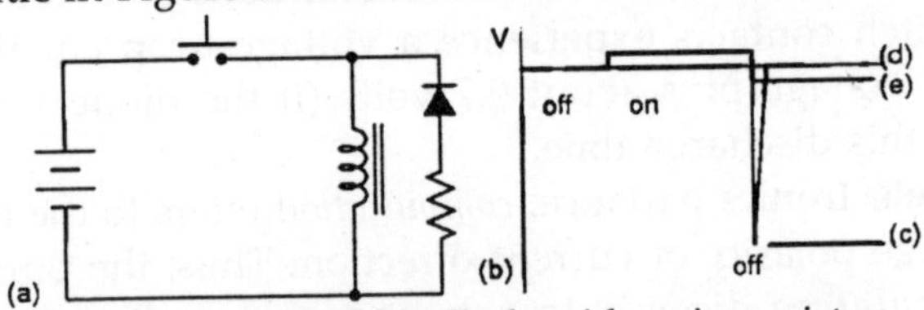

Fig. (a) Commutating diode with series resistor. (b) Voltage waveform. (c) Level with no diode. (d) Level with diode, no resistor. (e) Compromise level with diode and resistor.

A resistor placed in series with the commutating diode allows the coil's induced voltage to rise to a level greater than the diode's forward voltage drop, thus hastening the process of de-magnetization.

This, of course, will place the switch contacts under greater stress, and so the resistor must be sized to limit that transient voltage at an acceptable maximum level.

DIODE SWITCHING CIRCUITS

Diodes can perform switching and digital logic operations. Forward and reverse bias switch a diode between the low and high impedance states, respectively. Thus, it serves as a switch.

Logic

Diodes can perform digital logic functions: AND, and OR. Diode logic was used in early digital computers. It only finds limited application today. Sometimes it is convenient to fashion a single logic gate from a few diodes.

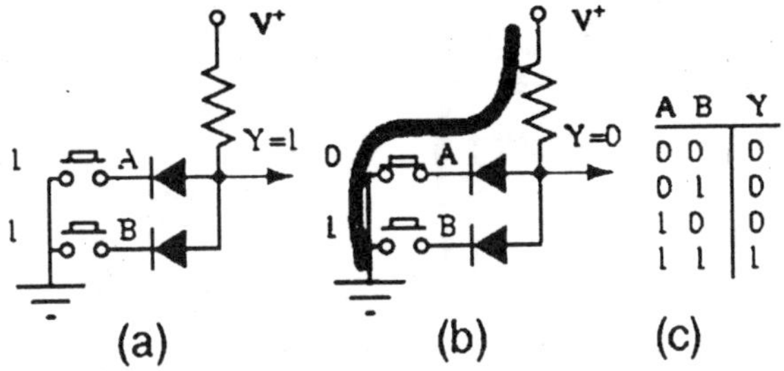

Fig. *Diode AND Gate*

An AND gate is shown in Figure, logic gates have inputs and an output (Y) which is a function of the inputs. The inputs to the gate are high (logic 1), say 10 V, or low, 0 V (logic 0). In the figure, the logic levels are generated by switches. If a switch is up, the input is effectively high (1). If the switch is down, it connects a diode cathode to ground, which is low (0). The output depends on the combination of inputs at A and B. The inputs and output are customarily recorded in a "truth table" at (c) to describe the logic of a gate.

At (a) all inputs are high (1). This is recorded in the last line of the truth table at (c). The output, Y, is high (1) due to the V^+ on the top of the resistor. It is unaffected by open switches. At (b) switch A pulls the cathode of the connected diode low, pulling output Y low (0.7 V). This is recorded in the third line of the truth table. The second line of the truth table describes the output with the switches reversed from (b). Switch B pulls the diode and output low.

The first line of the truth table recordes the Output=0 for both input low (0). The truth table describes a logical AND function.

A two input OR gate composed of a pair of diodes is shown in Figure. If both inputs are logic low at (a) as simulated by both switches "downward," the output Y is pulled low by the resistor. This logic zero is recorded in the first line of the truth table at (c). If one of the inputs is high as at (b), or the other input is high, or both inputs high, the diode(s) conduct(s), pulling the output Y high. These results are reordered in the second through fourth lines of the truth table. Summary: any input "high" is a high out at Y.

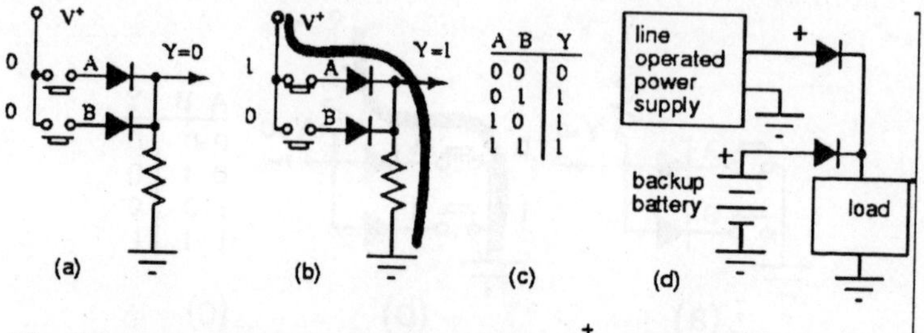

Fig. OR gate: (a) First line, Truth Table (TT). (b) Third line TT. (d) Logical OR of Power line Supply and Back-up Battery.

A backup battery may be OR-wired with a line operated DC power supply in Figure (d) to power a load, even during a power failure.

With AC power present, the line supply powers the load, assuming that it is a higher voltage than the battery. In the event of a power failure, the line supply voltage drops to 0 V; the battery powers the load.

The diodes must be is series with the power sources to prevent a failed line supply from draining the battery, and to prevent it from over charging the battery when line power is available.

Does your PC computer retain its BIOS setting when powered off? Does your VCR (video cassette recorder) retain the clock setting after a power failure? (PC Yes, old VCR no, new VCR yes.)

Analog switch: Diodes can switch analog signals. A reverse biased diode appears to be an open circuit. A forward biased diode is a low resistance conductor. The only problem is isolating the AC signal being switched from the DC control signal.

The circuit in Figure is a parallel resonant network: resonant tuning inductor paralleled by one (or more) of the switched resonator capacitors. This parallel LC resonant circuit could be a preselector filter for a radio receiver. It could be the frequency determining network of an oscillator (not shown). The digital control lines may be driven by a microprocessor interface.

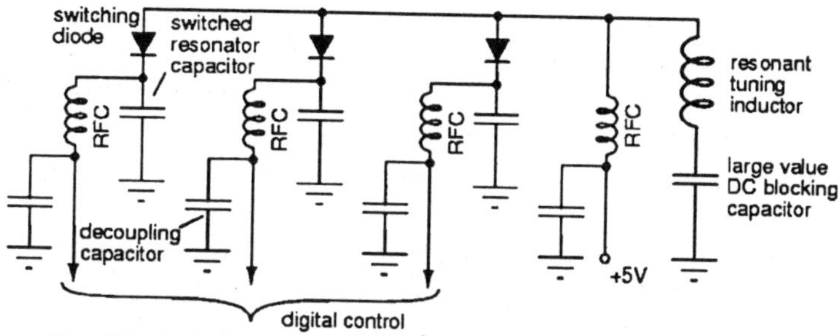

Fig. Diode Switch: A Digital Control Signal (low) Selects a Resonator Capacitor by Forward Biasing the Switching Diode.

The large value DC blocking capacitor grounds the resonant tuning inductor for AC while blocking DC. It would have a low reactance compared to the parallel LC reactances. This prevents the anode DC voltage from being shorted to ground by the resonant tuning inductor. A switched resonator capacitor is selected by pulling the corresponding digital control low. This forward biases the switching diode. The DC current path is from +5 V through an RF choke (RFC), a switching diode, and an RFC to ground via the digital control. The purpose of the RFC at the +5 V is to keep AC out of the +5 V supply. The RFC in series with the digital control is to keep AC out of the external control line. The decoupling capacitor shorts the little AC leaking through the RFC to ground, bypassing the external digital control line.

With all three digital control lines high (≥ + 5 V), no switched resonator capacitors are selected due to diode reverse bias. Pulling one or more lines low, selects one or more switched resonator capacitors, respectively. As more capacitors are switched in parallel with the resonant tuning inductor, the resonant frequency decreases.

The reverse biased diode capacitance may be substantial compared with very high frequency or ultra high frequency circuits. PIN diodes may be used as switches for lower capacitance.

ZENER DIODES

If we connect a diode and resistor in series with a DC voltage source so that the diode is forward-biased, the voltage drop across the diode will remain fairly constant over a wide range of power supply voltages as in Figure (a).

According to the "diode equation", the current through a forward-biased PN junction is proportional to *e* raised to the power of the forward voltage drop. Because this is an exponential function, current rises quite rapidly for modest increases in voltage drop. Another way of considering this is to say that voltage dropped across a forward-biased diode changes little for large variations in diode current. In the circuit shown above, diode current is limited by the voltage of the power supply.

The series resistor, and the diode's voltage drop, which as we know doesn't vary much from 0.7 volts. If the power supply voltage were to be increased, the resistor's voltage drop would increase almost the same amount, and the diode's voltage drop just a little.

Conversely, a decrease in power supply voltage would result in an almost equal decrease in resistor voltage drop, with just a little decrease in diode voltage drop. In a word, we could summarize this behavior by saying that the diode is *regulating* the voltage drop at approximately 0.7 volts.

Voltage regulation is a useful diode property to exploit. Suppose we were building some kind of circuit which could not tolerate variations in power supply voltage, but needed to be powered by a chemical battery, whose voltage changes over its lifetime.

We could form a circuit as shown and connect the circuit requiring steady voltage across the diode, where it would receive an unchanging 0.7 volts.

This would certainly work, but most practical circuits of any kind require a power supply voltage in excess of 0.7 volts to properly function.

One way we could increase our voltage regulation point would be to connect multiple diodes in series, so that their individual forward voltage drops of 0.7 volts each would add to create a larger total. For instance, if we had ten diodes in series, the regulated voltage would be ten times 0.7, or 7 volts in Figure (b).

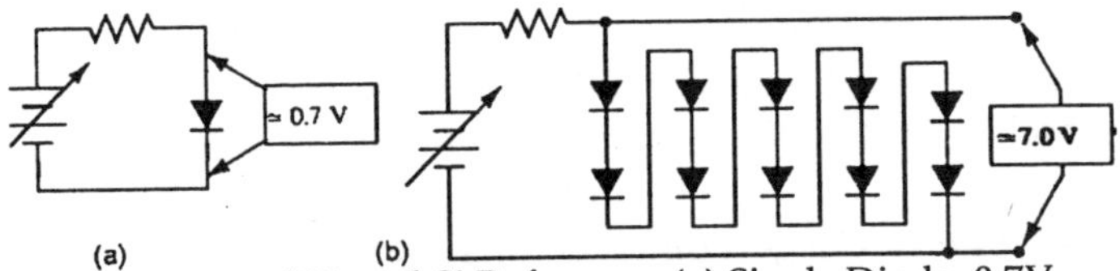

Fig. Forward Biased Si Reference: (a) Single Diode, 0.7V, (b) 10-Diodes in Series 7.0V.

So long as the battery voltage never sagged below 7 volts, there would always be about 7 volts dropped across the ten-diode "stack."

If larger regulated voltages are required, we could either use more diodes in series (an inelegant option, in my opinion), or try a fundamentally different approach. We know that diode forward voltage is a fairly constant figure under a wide range of conditions, but so is *reverse breakdown voltage*, and breakdown voltage is typically much, much greater than forward voltage. If we reversed the polarity of the diode in our single-diode regulator circuit and increased the power supply voltage to the point where the diode "broke down" (could no longer withstand the reverse-bias voltage impressed across it), the diode would similarly regulate the voltage at that breakdown point, not allowing it to increase further as in Figure below (a).

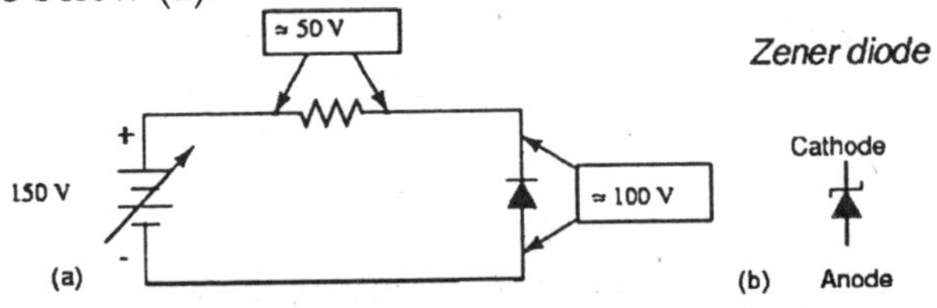

Fig. (a) Reverse Biased Si Small-signal Diode Breaks Down at about 100V. (b) Symbol for Zener Diode.

Unfortunately, when normal rectifying diodes "break down," they usually do so destructively. However, it is possible to build a special type of diode that can handle breakdown without failing completely. This type of diode is called a *zener diode*, and its symbol looks like Figure above (b).

When forward-biased, zener diodes behave much the same as standard rectifying diodes: they have a forward voltage drop which follows the "diode equation" and is about

0.7 volts. In reverse-bias mode, they do not conduct until the applied voltage reaches or exceeds the so-called *zener voltage*, at which point the diode is able to conduct substantial current, and in doing so will try to limit the voltage dropped across it to that zener voltage point. So long as the power dissipated by this reverse current does not exceed the diode's thermal limits, the diode will not be harmed.

Zener diodes are manufactured with zener voltages ranging anywhere from a few volts to hundreds of volts. This zener voltage changes slightly with temperature, and like common carbon-composition resistor values, may be anywhere from 5 percent to 10 percent in error from the manufacturer's specifications. However, this stability and accuracy is generally good enough for the zener diode to be used as a voltage regulator device in common power supply circuit in Figure.

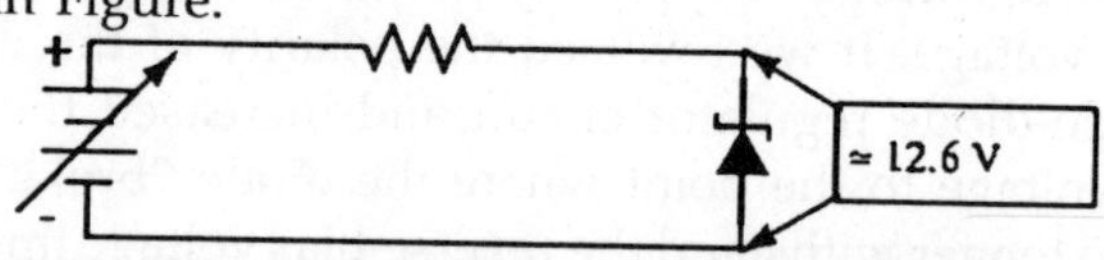

Fig. Zener Diode Regulator Circuit, Zener voltage = *12.6V).*

Please take note of the zener diode's orientation in the above circuit: the diode is *reverse-biased*, and intentionally so. If we had oriented the diode in the "normal" way, so as to be forward-biased, it would only drop 0.7 volts, just like a regular rectifying diode. If we want to exploit this diode's reverse breakdown properties, we must operate it in its reverse-bias mode. So long as the power supply voltage remains above the zener voltage (12.6 volts, in this example), the voltage dropped across the zener diode will remain at approximately 12.6 volts.

Like any semiconductor device, the zener diode is sensitive to temperature. Excessive temperature will destroy a zener diode, and because it both drops voltage and conducts current, it produces its own heat in accordance with Joule's Law (P=IE). Therefore, one must be careful to design the regulator circuit in such a way that the diode's power dissipation rating is not exceeded. Interestingly enough, when zener diodes fail due to excessive power dissipation, they usually fail *shorted* rather than open. A diode failed in this manner is readily detected: it drops almost zero voltage when biased either way, like a piece of wire.

Let's examine a zener diode regulating circuit mathematically, determining all voltages, currents, and power dissipations. Taking the same form of circuit shown earlier, we'll perform calculations assuming a zener voltage of 12.6 volts, a power supply voltage of 45 volts, and a series resistor value of 1000 Ω (we'll regard the zener voltage to be *exactly* 12.6 volts so as to avoid having to qualify all figures as "approximate" in Figure (a)

If the zener diode's voltage is 12.6 volts and the power supply's voltage is 45 volts, there will be 32.4 volts dropped across the resistor (45 volts - 12.6 volts = 32.4 volts). 32.4 volts dropped across 1000 Ω gives 32.4 mA of current in the circuit.

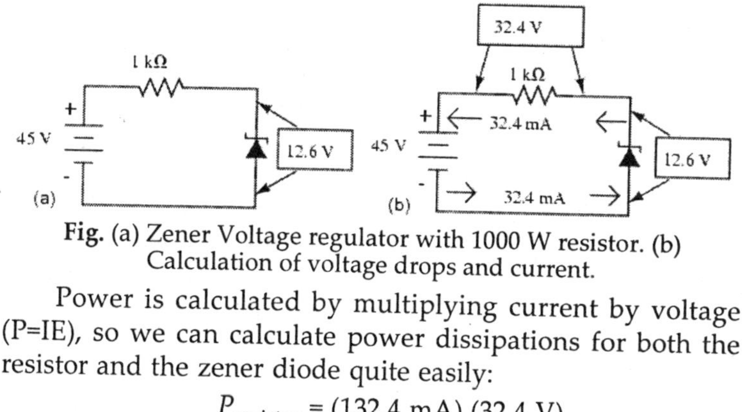

Fig. (a) Zener Voltage regulator with 1000 W resistor. (b) Calculation of voltage drops and current.

Power is calculated by multiplying current by voltage (P=IE), so we can calculate power dissipations for both the resistor and the zener diode quite easily:

$$P_{\text{resistor}} = (132.4 \text{ mA})\ (32.4 \text{ V})$$
$$P_{\text{resistor}} = 1.0498 \text{ W}$$
$$P_{\text{diode}} = (32.4 \text{ mA})(12.6 \text{ V})$$
$$P_{\text{diode}} = 408.24 \text{ mW}.$$

A zener diode with a power rating of 0.5 watt would be adequate, as would a resistor rated for 1.5 or 2 watts of dissipation.

If excessive power dissipation is detrimental, then why not design the circuit for the least amount of dissipation possible? Why not just size the resistor for a very high value of resistance, thus severely limiting current and keeping power dissipation figures very low? Take this circuit, for example, with a 100 kΩ resistor instead of a 1 kΩ resistor. Note that both the power supply voltage and the diode's zener voltage in Figure are identical to the last example:

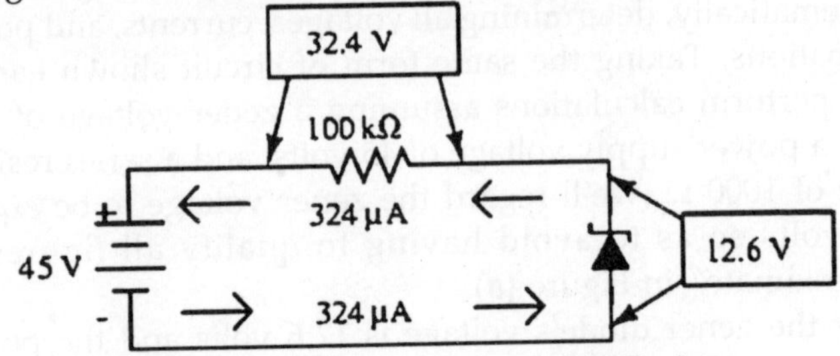

Fig. Zener Regulator with 100 kW Resistor

With only 1/100 of the current we had before (324 μA instead of 32.4 mA), both power dissipation figures should be 100 times smaller:

$$P_{resistor} = (324\ \mu A)\ (32.4\ V)$$
$$P_{resistor} = 10.498\ mW$$
$$P_{diode} = (324\ \mu A)(12.6\ V)$$
$$P_{diode} = 4.0824\ mW.$$

Seems ideal, doesn't it? Less power dissipation means lower operating temperatures for both the diode and the resistor, and also less wasted energy in the system, right? A higher resistance value *does* reduce power dissipation levels in the circuit, but it unfortunately introduces another problem. Remember that the purpose of a regulator circuit is to provide a stable voltage *for another circuit*. In other words, we're eventually going to power something with 12.6 volts, and this something will have a current draw of its own. Consider our first regulator circuit, this time with a 500 Ω load connected in parallel with the zener diode in Figure below.

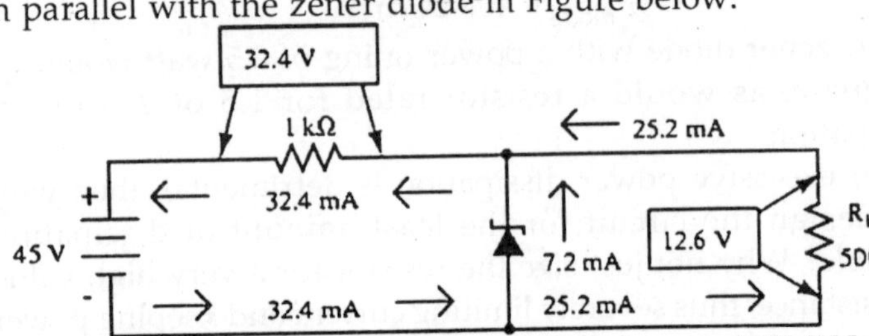

Fig. Zener Regulator with 1000 W Series Resistor and 500 W Load.

If 12.6 volts is maintained across a 500 Ω load, the load will draw 25.2 mA of current. In order for the 1 kΩ series "dropping" resistor to drop 32.4 volts (reducing the power supply's voltage of 45 volts down to 12.6 across the zener), it still must conduct 32.4 mA of current. This leaves 7.2 mA of current through the zener diode.

Now consider our "power-saving" regulator circuit with the 100 kΩ dropping resistor, delivering power to the same 500 Ω load. What it is supposed to do is maintain 12.6 volts across the load, just like the last circuit. However, as we will see, it *cannot* accomplish this task.

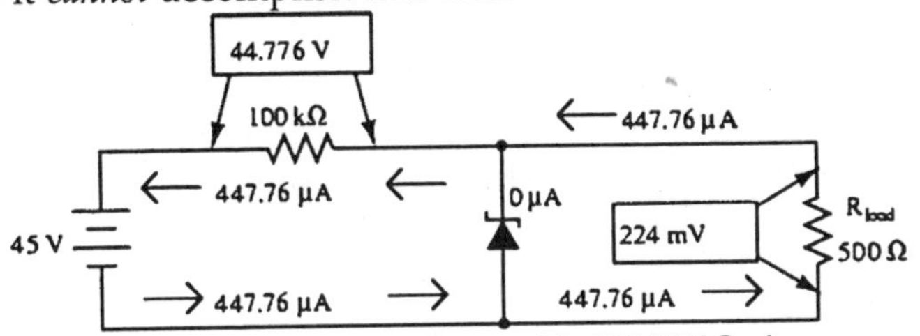

Fig. Zener non-Regulator with 100 KW Series Resistor with 500 W Load.>

With the larger value of dropping resistor in place, there will only be about 224 mV of voltage across the 500 Ω load, far less than the expected value of 12.6 volts! Why is this? If we actually had 12.6 volts across the load, it would draw 25.2 mA of current, as before.

This load current would have to go through the series dropping resistor as it did before, but with a new (much larger!) dropping resistor in place, the voltage dropped across that resistor with 25.2 mA of current going through it would be 2,520 volts! Since we obviously don't have that much voltage supplied by the battery, this cannot happen.

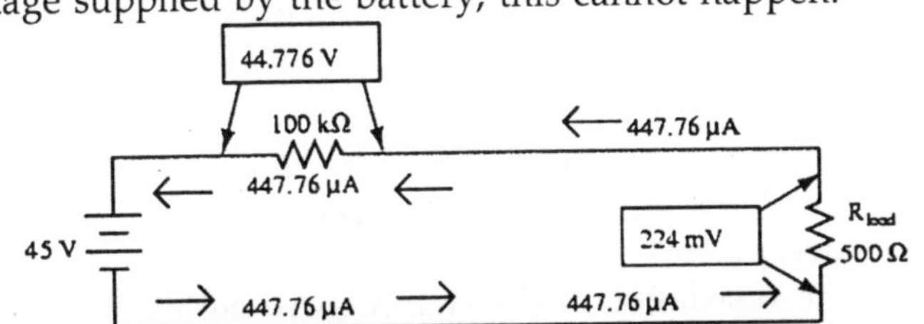

Fig. Non-Regulator with Zener Removed

The situation is easier to comprehend if we temporarily remove the zener diode from the circuit and analyze the behavior of the two resistors alone in Figure above.

Both the 100 kΩ dropping resistor and the 500 Ω load resistance are in series with each other, giving a total circuit resistance of 100.5 kΩ. With a total voltage of 45 volts and a total resistance of 100.5 kΩ, Ohm's Law (I=E/R) tells us that the current will be 447.76 μA. Figuring voltage drops across both resistors (E=IR), we arrive at 44.776 volts and 224 mV, respectively. If we were to re-install the zener diode at this point, it would "see" 224 mV across it as well, being in parallel with the load resistance.

This is far below the zener breakdown voltage of the diode and so it will not "break down" and conduct current. For that matter, at this low voltage the diode wouldn't conduct even if it were forward-biased! Thus, the diode ceases to regulate voltage. At least 12.6 volts must be dropped across to "activate" it.

The analytical technique of removing a zener diode from a circuit and seeing whether or not enough voltage is present to make it conduct is a sound one. Just because a zener diode happens to be connected in a circuit doesn't guarantee that the full zener voltage will always be dropped across it! Remember that zener diodes work by *limiting* voltage to some maximum level; they cannot *make up* for a lack of voltage.

In summary, any zener diode regulating circuit will function so long as the load's resistance is equal to or greater than some minimum value. If the load resistance is too low, it will draw too much current, dropping too much voltage across the series dropping resistor, leaving insufficient voltage across the zener diode to make it conduct. When the zener diode stops conducting current, it can no longer regulate voltage, and the load voltage will fall below the regulation point.

Our regulator circuit with the 100 kΩ dropping resistor must be good for some value of load resistance, though. To find this acceptable load resistance value, we can use a table to calculate resistance in the two-resistor series circuit (no diode), inserting the known values of total voltage and dropping resistor resistance, and calculating for an expected load voltage of 12.6 volts:

	$R_{dropping}$	R_{load}	Total	
E		12.6	45	Volts
I				Amps
R	100 k			Ohms

With 45 volts of total voltage and 12.6 volts across the load, we should have 32.4 volts across $R_{dropping}$:

	$R_{dropping}$	R_{load}	Total	
E	**32.4**	12.6	45	Volts
I				Amps
R	100 k			Ohms

With 32.4 volts across the dropping resistor, and 100 kΩ worth of resistance in it, the current through it will be 324 μA:

	$R_{dropping}$	R_{load}	Total	
E	32.4	12.6	45	Volts
I	**324 μ**			Amps
R	100 k			Ohms

Ohm's Law

$$I = \frac{E}{R}$$

Being a series circuit, the current is equal through all components at any given time:

	$R_{dropping}$	R_{load}	Total	
E	32.4	12.6	45	Volts
I	324 μ	**324 μ**	**324 μ**	Amps
R	100 k			Ohms

Rule of series circuits:

$$I_{Total} = I_1 = I_2 = \ldots I_n$$

Calculating load resistance is now a simple matter of Ohm's Law (R = E/I), giving us 38.889 kΩ:

	$R_{dropping}$	R_{load}	Total	
E	32.4	12.6	45	Volts
I	324 μ	324 μ	324 μ	Amps
R	100 k	**38.889 k**		Ohms

Ohm's Law

$$R = \frac{E}{I}$$

Thus, if the load resistance is exactly 38.889 kΩ, there will be 12.6 volts across it, diode or no diode. Any load resistance smaller than 38.889 kΩ will result in a load voltage less than 12.6 volts, diode or no diode. With the diode in place, the load voltage will be regulated to a maximum of 12.6 volts for any load resistance *greater* than 38.889 kΩ.

With the original value of 1 kΩ for the dropping resistor, our regulator circuit was able to adequately regulate voltage even for a load resistance as low as 500 Ω. What we see is a tradeoff between power dissipation and acceptable load resistance.

The higher-value dropping resistor gave us less power dissipation, at the expense of raising the acceptable minimum load resistance value. If we wish to regulate voltage for low-value load resistances, the circuit must be prepared to handle higher power dissipation.

Zener diodes regulate voltage by acting as complementary loads, drawing more or less current as necessary to ensure a constant voltage drop across the load.

This is analogous to regulating the speed of an automobile by braking rather than by varying the throttle position: not only is it wasteful, but the brakes must be built to handle all the engine's power when the driving conditions don't demand it.

Despite this fundamental inefficiency of design, zener diode regulator circuits are widely employed due to their sheer simplicity. In high-power applications where the inefficiencies would be unacceptable, other voltage-regulating techniques are applied.

But even then, small zener-based circuits are often used to provide a "reference" voltage to drive a more efficient amplifier circuit controlling the main power.

Zener diodes are manufactured in standard voltage ratings listed in Table below.

The table "Common zener diode voltages" lists common voltages for 0.3W and 1.3W parts. The wattage corresponds to die and package size, and is the power that the diode may dissipate without damage.

Common zener diode voltages

0.5W						
2.7V	3.0V	3.3V	3.6V	3.9V	4.3V	4.7V
5.1V	5.6V	6.2V	6.8V	7.5V	8.2V	9.1V
10V	11V	12V	13V	15V	16V	18V
20V	24V	27V	30V–			
1.3W						
4.7V	5.1V	5.6V	6.2V	6.8V	7.5V	8.2V
9.1V	10V	11V	12V	13V	15V	16V
18V	20V	22V	24V	27V	30V	33V
36V	39V	43V	47V	51V	56V	62V
68V	75V	100V	200V–			

SPECIAL-PURPOSE DIODES

Schottky Diodes

Schottky diodes are constructed of a *metal*-to-N junction rather than a P-N semiconductor junction. Also known as *hot-carrier* diodes, Schottky diodes are characterized by fast switching times (low reverse-recovery time), low forward voltage drop (typically 0.25 to 0.4 volts for a metal-silicon junction), and low junction capacitance.

The schematic symbol for a schottky diode is shown below.

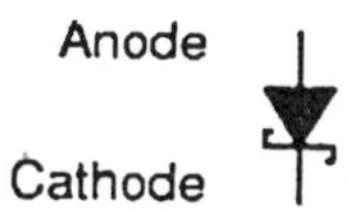

Fig. Schottky Diode Schematic Symbol.

The forward voltage drop (V_F), reverse-recovery time (t_{rr}), and junction capacitance (C_J) of Schottky diodes are closer to ideal than the average "rectifying" diode. This makes them well suited for high-frequency applications. Unfortunately, though, Schottky diodes typically have lower forward current (I_F) and reverse voltage (V_{RRM} and V_{DC}) ratings than rectifying diodes and are thus unsuitable for applications involving substantial amounts of power. Though they are used in low voltage switching regulator power supplies.

Schottky diode technology finds broad application in high-speed computer circuits, where the fast switching time equates to high speed capability, and the low forward voltage drop equates to less power dissipation when conducting.

Switching regulator power supplies operating at 100's of kHz cannot use conventional silicon diodes as rectifiers because of their slow switching speed . When the signal applied to a diode changes from forward to reverse bias, conduction continues for a short time, while carriers are being swept out of the depletion region. Conduction only ceases after this t_r *reverse recovery time* has expired. Schottky diodes have a shorter reverse recovery time.

Regardless of switching speed, the 0.7 V forward voltage drop of silicon diodes causes poor efficiency in low voltage supplies. This is not a problem in, say, a 10 V supply. In a 1 V supply the 0.7 V drop is a substantial portion of the output. One solution is to use a schottky power diode which has a lower forward drop.

Tunnel diodes

Tunnel diodes exploit a strange quantum phenomenon called *resonant tunneling* to provide a negative resistance forward-bias characteristics. When a small forward-bias voltage is applied across a tunnel diode, it begins to conduct current.

As the voltage is increased, the current increases and reaches a peak value called the *peak current* (I_P). If the voltage is increased a little more, the current actually begins to *decrease* until it reaches a low point called the *valley current* (I_V). If the voltage is increased further yet, the current begins to increase again, this time without decreasing into another "valley." The schematic symbol for the tunnel diode .

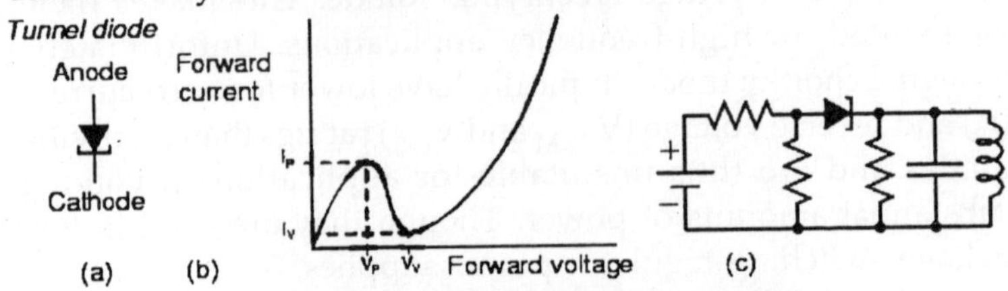

Fig. Tunnel Diode (a) Schematic Symbol. (b) Current vs Voltage plot (c) Oscillator.

The forward voltages necessary to drive a tunnel diode to its peak and valley currents are known as peak voltage (V_P) and valley voltage (V_V), respectively. The region on the graph where current is decreasing while applied voltage is increasing (between V_P and V_V on the horizontal scale) is known as the region of *negative resistance.*

Tunnel diodes, also known as *Esaki diodes* in honor of their Japanese inventor Leo Esaki, are able to transition between peak and valley current levels very quickly, "switching" between high and low states of conduction much faster than even Schottky diodes. Tunnel diode characteristics are also relatively unaffected by changes in temperature.

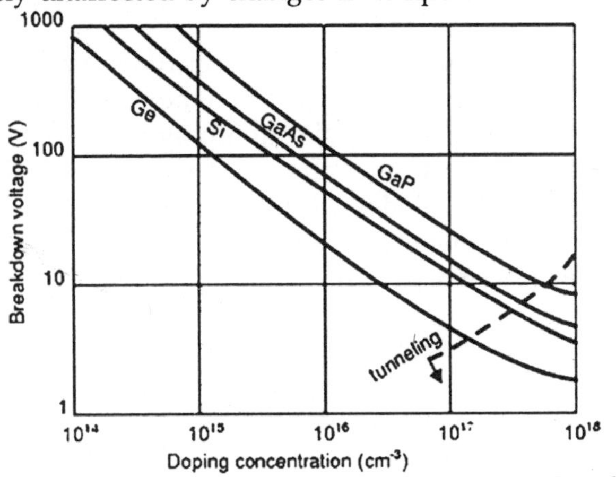

Fig. Reverse Breakdown Voltage Versus Doping Level.

Tunnel diodes are heavily doped in both the P and N regions, 1000 times the level in a rectifier. This can be seen in Figure above. Standard diodes are to the far left, zener diodes near to the left, and tunnel diodes to the right of the dashed line. The heavy doping produces an unusually thin depletion region. This produces an unusually low reverse breakdown voltage with high leakage. The thin depletion region causes high capacitance. To overcome this, the tunnel diode junction area must be tiny. The forward diode characteristic consists of two regions: a normal forward diode characteristic with current rising exponentially beyond V_F, 0.3 V for Ge, 0.7 V for Si. Between 0 V and V_F is an additional "negative resistance" characteristic peak. This is due to quantum mechanical tunneling involving the dual particle-wave nature of electrons. The depletion region is thin enough compared with the equivalent wavelength of the electron that they can tunnel through.

They do not have to overcome the normal forward diode voltage V_F. The energy level of the conduction band of the N-type material overlaps the level of the valence band in the P-type region. With increasing voltage, tunneling begins; the levels overlap; current increases, up to a point. As current increases further, the energy levels overlap less; current decreases with increasing voltage. This is the "negative resistance" portion of the curve.

Tunnel diodes are not good rectifiers, as they have relatively high "leakage" current when reverse-biased. Consequently, they find application only in special circuits where their unique tunnel effect has value. To exploit the tunnel effect, these diodes are maintained at a bias voltage somewhere between the peak and valley voltage levels, always in a forward-biased polarity (anode positive, and cathode negative).

Perhaps the most common application of a tunnel diode is in simple high-frequency oscillator circuits as in Figure above(c), where it allows a DC voltage source to contribute power to an LC "tank" circuit, the diode conducting when the voltage across it reaches the peak (tunnel) level and effectively insulating at all other voltages. The resistors bias the tunnel diode at a few tenths of a volt centered on the negative resistance portion of the characteristic curve. The L-C resonant circuit may be a section of waveguide for microwave operation. Oscillation to 5 GHz is possible.

At one time the tunnel diode was the only solid-state microwave amplifier available. Tunnel diodes were popular starting in the 1960's. They were longer lived than traveling wave tube amplifiers, an important consideration in satellite transmitters. Tunnel diodes are also resistant to radiation because of the heavy doping. Today various transistors operate at microwave frequencies. Even small signal tunnel diodes are expensive and difficult to find today. There is one remaining manufacturer of germanium tunnel diodes, and none for silicon devices. They are sometimes used in military equipment because they are insensitive to radiation and large temperature changes.

There has been some research involving possible integration of silicon tunnel diodes into CMOS integrated circuits. They are thought to be capable of switching at 100 GHz in digital circuits. The sole manufacturer of germanium devices produces them one at a time. A batch process for silicon tunnel diodes must be developed, then integrated with conventional CMOS processes.

The Esaki tunnel diode should not be confused with the *resonant tunneling diode* CH 2, of more complex construction from compound semiconductors. The RTD is a more recent development capable of higher speed.

Light-Emitting Diodes

Diodes, like all semiconductor devices, are governed by the principles described in quantum physics. One of these principles is the emission of specific-frequency radiant energy whenever electrons fall from a higher energy level to a lower energy level. This is the same principle at work in a neon lamp, the characteristic pink-orange glow of ionized neon due to the specific energy transitions of its electrons in the midst of an electric current. The unique color of a neon lamp's glow is due to the fact that its *neon* gas inside the tube, and not due to the particular amount of current through the tube or voltage between the two electrodes. Neon gas glows pinkish-orange over a wide range of ionizing voltages and currents. Each chemical element has its own "signature" emission of radiant energy when its electrons "jump" between different, quantized energy levels. Hydrogen gas, for example, glows red when ionized; mercury vapor glows blue. This is what makes spectrographic identification of elements possible.

Electrons flowing through a PN junction experience similar transitions in energy level, and emit radiant energy as they do so. The frequency of this radiant energy is determined by the crystal structure of the semiconductor material, and the elements comprising it. Some semiconductor junctions, composed of special chemical combinations, emit radiant energy within the spectrum of visible light as the electrons change energy levels. Simply put, these junctions *glow* when forward biased. A diode intentionally designed to glow like a lamp is called a *light-emitting diode*, or *LED*.

Forward biased silicon diodes give off heat as electron and holes from the N-type and P-type regions, respectively, recombine at the junction. In a forward biased LED, the recombination of electrons and holes in the active region in Figure below (c) yields photons. This process is known as *electroluminescence*. To give off photons, the potential barrier through which the electrons fall must be higher than for a silicon diode. The forward diode drop can range to a few volts for some color LEDs. Diodes made from a combination of the elements gallium, arsenic, and phosphorus (called *gallium-arsenide-phosphide*) glow bright red, and are some of the most common LEDs manufactured.

By altering the chemical constituency of the PN junction, different colors may be obtained. Some of the currently available colors other than red are green, blue, and infra-red (invisible light at a frequency lower than red). Other colors may be obtained by combining two or more primary-color (red, green, and blue) LEDs together in the same package, sharing the same optical lens. For instance, a yellow LED may be made by merging a red LED with a green LED. The schematic symbol for an LED is a regular diode shape inside of a circle, with two small arrows pointing away (indicating emitted light).

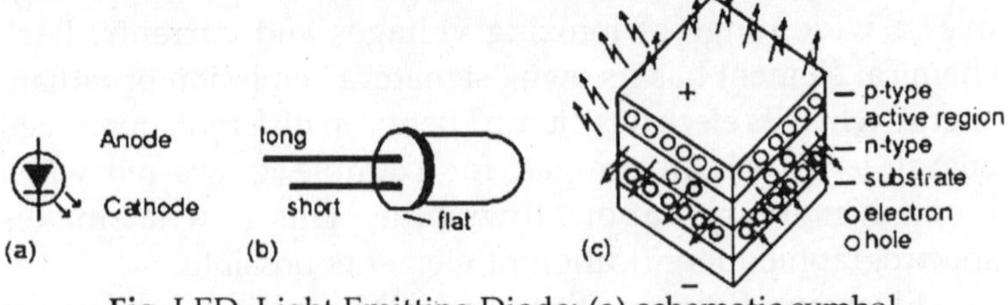

Fig. LED, Light Emitting Diode: (a) schematic symbol. (b) Flat side and short lead of device correspond to cathode. (c) Cross section of Led die.

This notation of having two small arrows pointing away from the device is common to the schematic symbols of all light-emitting semiconductor devices. Conversely, if a device is light-*activated* (meaning that incoming light stimulates it), then the symbol will have two small arrows pointing *toward* it. LEDs can sense light. They generate a small voltage when exposed to light, much like a solar cell on a small scale. This property can be gainfully applied in a variety of light-sensing circuits.

Because LEDs are made of different chemical substances than silicon diodes, their forward voltage drops will be different.

Typically, LEDs have much larger forward voltage drops than rectifying diodes, anywhere from about 1.6 volts to over 3 volts, depending on the color. Typical operating current for a standard-sized LED is around 20 mA. When operating an LED from a DC voltage source greater than the LED's forward voltage, a series-connected "dropping" resistor must be included to prevent full source voltage from damaging the LED. Consider the example circuit in Figure below (a) using a 6 V source.

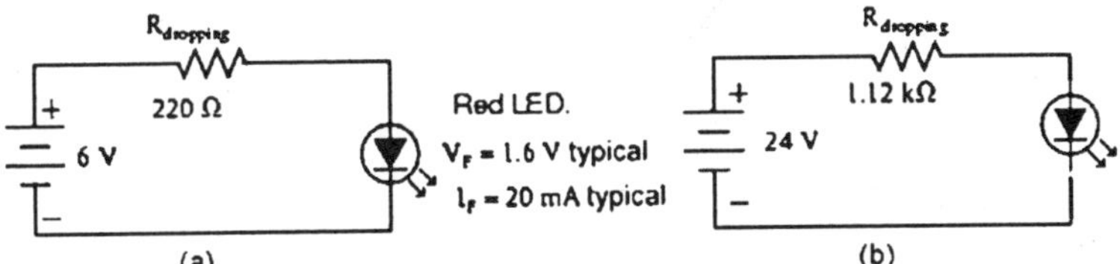

Fig. Setting LED Current at 20 ma.
(a) for a 6 V Source, (b) for a 24 V Source.

With the LED dropping 1.6 volts, there will be 4.4 volts dropped across the resistor. Sizing the resistor for an LED current of 20 mA is as simple as taking its voltage drop (4.4 volts) and dividing by circuit current (20 mA), in accordance with Ohm's Law (R=E/I). This gives us a figure of 220 Ω. Calculating power dissipation for this resistor, we take its voltage drop and multiply by its current (P=IE), and end up with 88 mW, well within the rating of a 1/8 watt resistor. Higher battery voltages will require larger-value dropping resistors, and possibly higher-power rating resistors as well.

Consider the example in Figure above (b) for a supply voltage of 24 volts:

Here, the dropping resistor must be increased to a size of 1.12 kΩ to drop 22.4 volts at 20 mA so that the LED still receives only 1.6 volts. This also makes for a higher resistor power dissipation: 448 mW, nearly one-half a watt of power! Obviously, a resistor rated for 1/8 watt power dissipation or even 1/4 watt dissipation will overheat if used here.

Dropping resistor values need not be precise for LED circuits. Suppose we were to use a 1 kΩ resistor instead of a 1.12 kΩ resistor in the circuit shown above. The result would be a slightly greater circuit current and LED voltage drop, resulting in a brighter light from the LED and slightly reduced service life. A dropping resistor with too much resistance (say, 1.5 kΩ instead of 1.12 kΩ) will result in less circuit current, less LED voltage, and a dimmer light. LEDs are quite tolerant of variation in applied power, so you need not strive for perfection in sizing the dropping resistor.

Multiple LEDs are sometimes required, say in lighting. If LEDs are operated in parallel, each must have its own current limiting resistor as in Figure below (a) to ensure currents dividing more equally. However, it is more efficient to operate LEDs in series with a single dropping resistor. As the number of series LEDs increases the series resistor value must decrease to maintain current, to a point. The number of LEDs in series (V_f) cannot exceed the capability of the power supply. Multiple series strings may be employed as in Figure below (c).

In spite of equalizing the currents in multiple LEDs, the brightness of the devices may not match due to variations in the individual parts. Parts can be selected for brightness matching for critical applications.

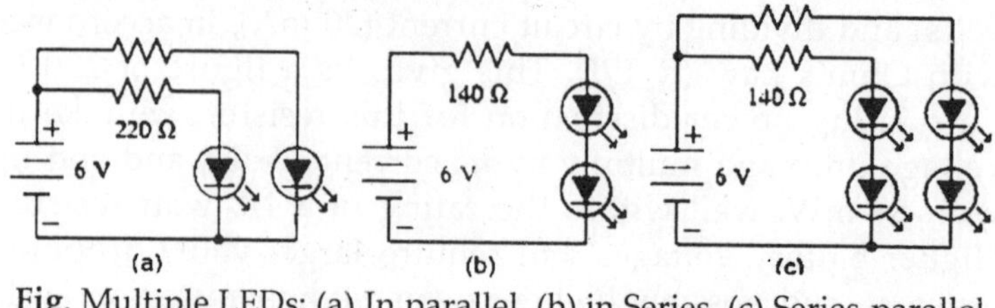

Fig. Multiple LEDs: (a) In parallel, (b) in Series, (c) Series-parallel

Also because of their unique chemical makeup, LEDs have much, much lower peak-inverse voltage (PIV) ratings than ordinary rectifying diodes. A typical LED might only be rated at 5 volts in reverse-bias mode. Therefore, when using alternating current to power an LED, connect a protective rectifying diode anti-parallel with the LED to prevent reverse breakdown every other half-cycle as in Figure below (a).

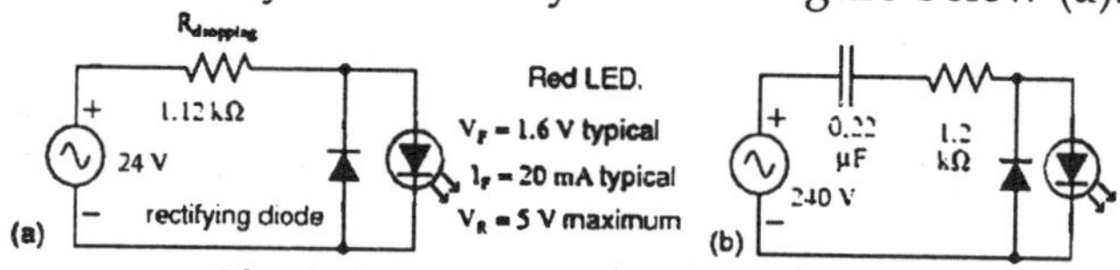

Fig. Safely driving an LED with AC:
(a) from 24 VAC, (b) from 240 VAC.

If the LED is driven from a 240 VAC source, the Figure above (a) voltage source is increased from 24 VAC to 240 VAC, the resistor from 1.12 kΩ to 12 kΩ. The power dissipated in the 12 kΩ resistor is an unattractive 4.8 watts.

$$P = VI = (240\ V)(20\ mA) = 4.8\ watt$$

A potential solution is to replace the 12 kΩ resistor with a non-dissipative 12 kΩ capacitive reactance. This would be Figure above (b) with the resistor shorted. That circuit at (b), missing the resistor, was published in an electrical engineering journal. This author constructed the circuit. It worked the first time it was powered "on," but not thereafter upon "power on". Each time it was powered "on," it got dimmer until it failed completely. Why? If "power on" occurs near a zero crossing of the AC sinewave, the circuit works.

However, if powered "on" at, say, the peak of the sinewave, the voltage rises abruptly from zero to the peak. Since the current through the capacitor is i = C(dv/dt), the current spikes to a very large value exceeding the "surge current" rating of the LED, destroying it.

The solution is to design a capacitor for the continuous current of the LED, and a series resistor to limit current during "power on" to the surge current rating of the LED. Often the surge current rating of an LED is ten times higher than the continuous current rating. (Though, this is not true of high current illumination grade LED's.) We calculate a capacitor to supply 20 mA continuous current, then select a resistor having resistance of 1/10 th the capacitive reactance.

$$I = 20 \text{ mA}$$
$$X_c = (240 \text{ V})/(20 \text{ mA}) = 12 \text{ kW}$$
$$X_c = 1/2Pf_c$$
$$C = 1/2PX_c = 1/2P60(12 \text{ kW} = 0.22 \ \mu\text{F}$$
$$R = (0.10)X_c = (0.10)(12\text{kW}) = 1.2 \text{ kW}$$
$$P = I^2R = (20 \text{ mA})^2(1.2 \text{ kW}) = 0.48 \text{ watt}$$

The resistor limits the LED current to 200 mA during the "power on" surge. Thereafter it passes 20 mA as limited by the capacitor.

The 1.2 kresistor dissipates 0.48 watts compared with 4.8 watts for the 12 kΩ resistor circuit.

What component values would be required to operate the circuit on 120 VAC? One solution is to use the 240 VAC circuit on 120 VAC with no change in component values, halving the LED continuous current to 10 mA. If operation at 20 mA is required, double the capacitor value and halve the resistor value.

The anti-parallel diodes in Figure above can be replaced with an anti-parallel LED. The resulting pair of anti-parallel LED's illuminate on alternating half-cycles of the AC sinewave. This configuration draws 20 ma, splitting it equally between the LED's on alternating AC half cycles. Each LED only receives 10 mA due to this sharing.

The same is true of the LED anti-parallel combination with a rectifier. The LED only receives 10 ma. If 20 mA was required for the LED(s), The capacitor value in μF could be doubled and the resistor halved.

The forward voltage drop of LED's is inversely proportional to the wavelength (ë). As wavelength decreases going from infrared to visible colors to ultraviolet, V_f increases. While this trend is most obvious in the various devices from a single manufacturer.

The voltage range for a particular color LED from various manufacturers varies. This range of voltages is shown in Table below.

Optical and electrical properties of LED's

LED	*λ nm (= 10 ⁹m)*	V_f*(from)*	V_f *(to)*
infrared	940	1.2	1.7
red	660	1.5	2.4
orange	602	620-2.1	2.2
yellow, green	560	595-1.7	2.8
white, blue, violet	—	3	4
ultraviolet	370	4.2	4.8

As lamps, LEDs are superior to incandescent bulbs in many ways. First and foremost is efficiency: LEDs output far more light power per watt of electrical input than an incandescent lamp.

This is a significant advantage if the circuit in question is battery-powered, efficiency translating to longer battery life. Second is the fact that LEDs are far more reliable, having a much greater service life than incandescent lamps. This is because LEDs are "cold" devices: they operate at much cooler temperatures than an incandescent lamp with a white-hot metal filament, susceptible to breakage from mechanical and thermal shock.

Third is the high speed at which LEDs may be turned on and off. This advantage is also due to the "cold" operation of LEDs: they don't have to overcome thermal inertia in transitioning from off to on or vice versa. For this reason, LEDs are used to transmit digital (on/off) information as pulses of light, conducted in empty space or through fiber-optic cable, at very high rates of speed (millions of pulses per second).

LEDs excel in monochromatic lighting applications like traffic signals and automotive tail lights. Incandescents are abysmal in this application since they require filtering, decreasing efficiency. LEDs do not require filtering.

One major disadvantage of using LEDs as sources of illumination is their monochromatic (single-color) emission. No one wants to read a book under the light of a red, green, or blue LED.

However, if used in combination, LED colors may be mixed for a more broad-spectrum glow. A new broad spectrum light source is the white LED. While small white panel indicators have been available for many years, illumination grade devices are still in development.

Efficiency of Lighting

Lamp type	*Efficiency lumen/watt*	*Life hrs*	*notes*
White LED	35	100,000	costly
White LED, future	100	100,000	R&D target
Incandescent	12	1000	inexpensive
Halogen	15-17	2000	high quality light
Compact fluorescent	50-100	10,000	cost effective
Sodium vapor, lp	70-200	20,000	outdoor
Mercury vapor	13-48	18,000	outdoor

A white LED is a blue LED exciting a phosphor which emits yellow light. The blue plus yellow approximates white light. The nature of the phosphor determines the characteristics of the light. A red phosphor may be added to improve the quality of the yellow plus blue mixture at the expense of efficiency. Table above compares white illumination LEDs to expected future devices and other conventional lamps. Efficiency is measured in lumens of light output per watt of input power. If the 50 lumens/watt device can be improved to 100 lumens/watt, white LEDs will be comparable to compact fluorescent lamps in efficiency.

Laser Diodes

The *laser diode* is a further development upon the regular light-emitting diode, or LED. The term "laser" itself is actually an acronym, despite the fact its often written in lower-case letters. "Laser" stands for Light Amplification by Stimulated Emission of Radiation, and refers to another strange quantum process whereby characteristic light emitted by electrons falling from high-level to low-level energy states in a material stimulate other electrons in a substance to make similar "jumps," the result being a synchronized output of light from the material.

This synchronization extends to the actual *phase* of the emitted light, so that all light waves emitted from a "lasing" material are not just the same frequency (color), but also the same phase as each other, so that they reinforce one another and are able to travel in a very tightly-confined, nondispersing beam. This is why laser light stays so remarkably focused over long distances: each and every light wave coming from the laser is in step with each other.

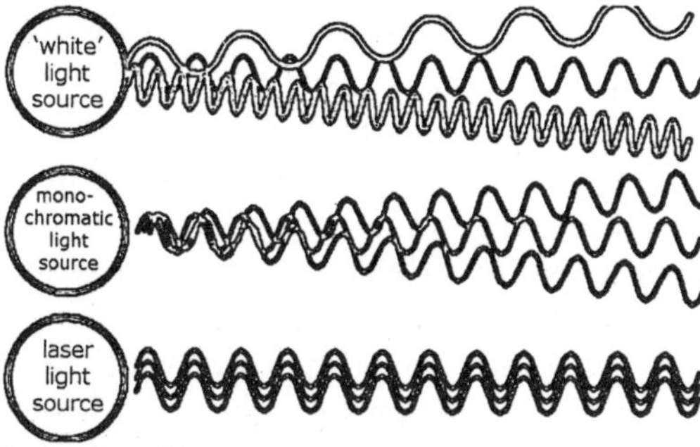

Fig (a) White Light of Many Wavelengths. (b) Mono-chromatic LED Light, A single Wavelength. (c) Phase Coherent Laser Light.

Incandescent lamps produce "white" (mixed-frequency, or mixed-color) light as in Figure above (a). Regular LEDs produce monochromatic light: same frequency (color), but different phases, resulting in similar beam dispersion in Figure above (b). Laser LEDs produce *coherent light*: light that is both monochromatic (single-color) and monophasic (single-phase), resulting in precise beam confinement as in Figure above (c).

Laser light finds wide application in the modern world: everything from surveying, where a straight and nondispersing light beam is very useful for precise sighting of measurement markers, to the reading and writing of optical disks, where only the narrowness of a focused laser beam is able to resolve the microscopic "pits" in the disk's surface comprising the binary 1's and 0's of digital information.

Some laser diodes require special high-power "pulsing" circuits to deliver large quantities of voltage and current in short bursts. Other laser diodes may be operated continuously at lower power. In the continuous laser, laser action occurs only within a certain range of diode current, necessitating some form of current-regulator circuit. As laser diodes age, their power requirements may change (more current required for less output power), but it should be remembered that low-power laser diodes, like LEDs, are fairly long-lived devices, with typical service lives in the tens of thousands of hours.

Photodiodes

A *photodiode* is a diode optimized to produce an electron current flow in response to irradiation by ultraviolet, visible, or infrared light. Silicon is most often used to fabricate photodiodes; though, germanium and gallium arsenide can be used. The junction through which light enters the semiconductor must be thin enough to pass most of the light on to the active region (depletion region) where light is converted to electron hole pairs.

In Figure below a shallow P-type diffusion into an N-type wafer produces a PN junction near the surface of the wafer. The P-type layer needs to be thin to pass as much light as possible. A heavy N+ diffusion on the back of the wafer makes contact with metalization.

The top metalization may be a fine grid of metallic fingers on the top of the wafer for large cells. In small photodiodes, the top contact might be a sole bond wire contacting the bare P-type silicon top.

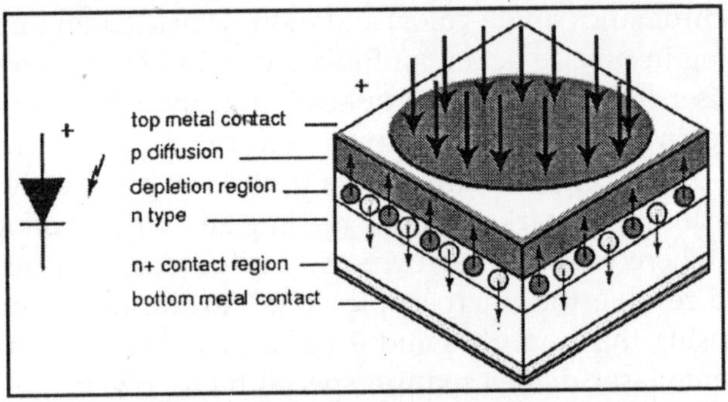

Fig. Photodiode: Schematic Symbol and Cross Section.

Light entering the top of the photodiode stack fall off exponentially in with depth of the silicon. A thin top P-type layer allows most photons to pass into the depletion region where electron-hole pairs are formed. The electric field across the depletion region due to the built in diode potential causes electrons to be swept into the N-layer, holes into the P-layer. Actually electron-hole pairs may be formed in any of the semiconductor regions. However, those formed in the depletion region are most likely to be separated into the respective N and P-regions.

Many of the electron-hole pairs formed in the P and N-regions recombine. Only a few do so in the depletion region. Thus, a few electron-hole pairs in the N and P-regions, and most in the depletion region contribute to *photocurrent,* that current resulting from light falling on the photodiode.

The voltage out of a photodiode may be observed. However, operation in this *photovoltaic (PV)* mode is not linear over a large dynamic range. Though, it is sensitive and low noise at low frequencies, < 100 kHz.

The preferred mode of operation is often *photocurrent (PC)* mode because the current is linearly proportional to light flux over several decades of intensity, and higher frequency response can be achieved. PC mode is achieved with reverse bias or zero bias on the photodiode.

A current amplifier (transimpedance amplifier) should be used with a photodiode in PC mode. Linearity and PC mode are achieved as long as the diode does not become forward biased.

High speed operation is often required of photodiodes, as opposed to solar cells. Speed is a function of diode capacitance, which can be minimized by decreasing cell area. Thus, a sensor for a high speed fiber optic link will use an area no larger than necessary, say 1 mm^2. Capacitance may also be decreased by increasing the thickness of the depletion region, in the manufacturing process or by increasing the reverse bias on the diode.

PIN diode

The *p-i-n diode* or *PIN diode* is a photodiode with an intrinsic layer between the P and N-regions as in Figure below. The P-Intrinsic-N structure increases the distance between the P and N conductive layers, decreasing capacitance, increasing speed.

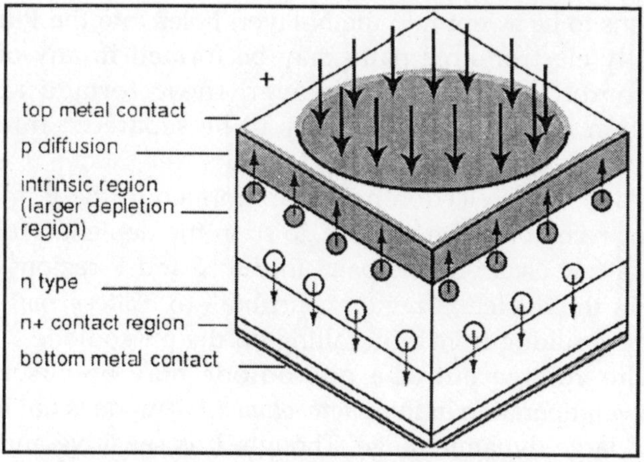

Fig. PIN photodiode: The Intrinsic Region Increases The Thickness of the Depletion Region.

Avalanche Photo Diode:

An *avalanche photodiode (APD)* designed to operate at high reverse bias exhibits an electron multiplier effect analogous to a photomultiplier tube. The reverse bias can run from 10's of volts to nearly 2000 V. The high level of reverse bias accelerates photon created electron-hole pairs in the intrinsic region to a high enough velocity to free additional carriers from collisions with the crystal lattice. Thus, many electrons per photon result. The motivation for the APD is to achieve amplification within the photodiode to overcome noise in external amplifiers.

This works to some extent. However, the APD creates noise of its own. At high speed the APD is superior to a PIN diode amplifier combination, though not for low speed applications. APD's are expensive, roughly the price of a photomultiplier tube. So, they are only competitive with PIN photodiodes for niche applications. One such application is single photon counting as applied to nuclear physics.

Solar Cells

A photodiode optimized for efficiently delivering power to a load is the *solar cell*. It operates in photovoltaic mode (PV) because it is forward biased by the voltage developed across the load resistance.

Monocrystalline solar cells are manufactured in a process similar to semiconductor processing. This involves growing a single crystal boule from molten high purity silicon (P-type), though, not as high purity as for semiconductors. The boule is diamond sawed or wire sawed into wafers. The ends of the boule must be discarded or recycled, and silicon is lost in the saw kerf. Since modern cells are nearly square, silicon is lost in squaring the boule. Cells may be etched to texture (roughen) the surface to help trap light within the cell. Considerable silicon is lost in producing the 10 or 15 cm square wafers. These days (2007) it is common for solar cell manufacturer to purchase the wafers at this stage from a supplier to the semiconductor industry.

P-type Wafers are loaded back-to-back into fused silica boats exposing only the outer surface to the N-type dopant in the diffusion furnace. The diffusion process forms a thin n-type layer on the top of the cell. The diffusion also shorts the edges of the cell front to back. The periphery must be removed by plasma etching to unshort the cell. Silver and or aluminum paste is screened on the back of the cell, and a silver grid on the front. These are sintered in a furnace for good electrical contact. The cells are wired in series with metal ribbons. For charging 12 V batteries, 36 cells at approximately 0.5 V are vacuum laminated between glass, and a polymer metal back. The glass may have a textured surface to help trap light.

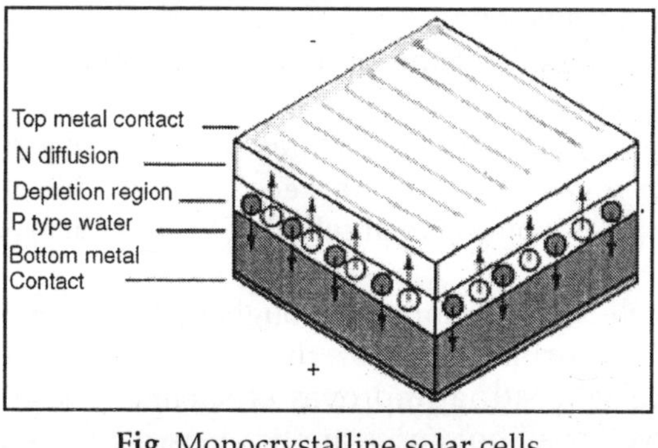

Fig. Monocrystalline solar cells

Silicon Solar Cell

The ultimate commercial high efficiency (21.5%) single crystal silicon solar cells have all contacts on the back of the cell. The active area of the cell is increased by moving the top (-) contact conductors to the back of the cell.

The top (-) contacts are normally made to the N-type silicon on top of the cell. In Figure below the (-) contacts are made to N^+ diffusions on the bottom interleaved with (+) contacts. The top surface is textured to aid in trapping light within the cell.. [VSW]

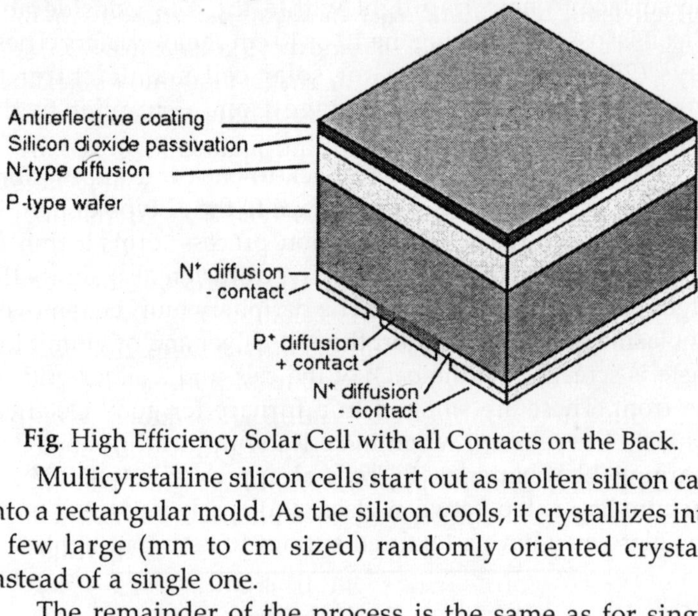

Fig. High Efficiency Solar Cell with all Contacts on the Back.

Multicyrstalline silicon cells start out as molten silicon cast into a rectangular mold. As the silicon cools, it crystallizes into a few large (mm to cm sized) randomly oriented crystals instead of a single one.

The remainder of the process is the same as for single crystal cells.

The finished cells show lines dividing the individual crystals, as if the cells were cracked. The high efficiency is not quite as high as single crystal cells due to losses at crystal grain boundaries.

The cell surface cannot be roughened by etching due to the random orientation of the crystals. However, an antireflectrive coating improves efficiency. These cells are competitive for all but space applications.

Three Layer Cell:

The highest efficiency solar cell is a stack of three cells tuned to absorb different portions of the solar spectrum. Though three cells can be stacked atop one another, a monolithic single crystal structure of 20 semiconductor layers is more compact. At 32 % efficiency, it is now (2007) favored over silicon for space application. The high cost prevents it from finding many earth bound applications other than concentrators based on lenses or mirrors.

Intensive research has recently produced a version enhanced for terrestrial concentrators at 400 - 1000 suns and 40.7% efficiency.

This requires either a big inexpensive Fresnel lens or reflector and a small area of the expensive semiconductor. This combination is thought to be competitive with inexpensive silicon cells for solar power plants.

Metal organic chemical vapor deposition (MOCVD) deposits the layers atop a P-type germanium substrate. The top layers of N and P-type gallium indium phosphide (GaInP) having a band gap of 1.85 eV, absorbs ultraviolet and visible light. These wavelengths have enough energy to exceed the band gap. Longer wavelengths (lower energy) do not have enough energy to create electron-hole pairs, and pass on through to the next layer.

A gallium arsenide layers having a band gap of 1.42 eV, absorbs near infrared light. Finally the germanium layer and substrate absorb far infrared. The series of three cells produce a voltage which is the sum of the voltages of the three cells. The voltage developed by each material is 0.4 V less than the band gap energy listed in Table below. For example, for GaInP: 1.8 eV/e - 0.4 V = 1.4 V. For all three the voltage is 1.4 V + 1.0 V + 0.3 V = 2.7 V.

High Efficiency Triple Layer Solar Cell.

Layer	*Band gap*	*Light absorbed*
Gallium indium phosphide	1.8 eV	UV, visible
Gallium arsenide	1.4 eV	near infrared
Germanium	0.7 eV	far infrared

Crystalline solar cell arrays have a long useable life. Many arrays are guaranteed for 25 years, and believed to be good for 40 years. They do not suffer initial degradation compared with amorphous silicon.

Both single and multicrystalline solar cells are based on silicon wafers. The silicon is both the substrate and the active device layers. Much silicon is consumed. This kind of cell has been around for decades, and takes approximately 86% of the solar electric market. For further information about crystalline solar cells see Honsberg.

Amorphous silicon thin film solar cells use tiny amounts of the active raw material, silicon. Approximately half the cost of conventional crystalline solar cells is the solar cell grade silicon. The thin film deposition process reduces this cost. The downside is that efficiency is about half that of conventional crystalline cells.

Moreover, efficiency degrades by 15-35% upon exposure to sunlight. A 7% efficient cell soon ages to 5% efficiency. Thin film amorphous silicon cells work better than crystalline cells in dim light. They are put to good use in solar powered calculators.

Non-silicon based solar cells make up about 7% of the market. These are thin-film polycrystalline products. Various compound semiconductors are the subject of research and development. Some non-silicon products are in production. Generally, the efficiency is better than amorphous silicon, but not nearly as good as crystalline silicon.

Cadmium telluride as a polycrystalline thin film on metal or glass can have a higher efficiency than amorphous silicon thin films. If deposited on metal, that layer is the negative contact to the cadmium telluride thin film. The transparent P-type cadmium sulfide atop the cadmium telluride serves as a buffer layer.

The positive top contact is transparent, electrically conductive fluorine doped tin oxide. These layers may be laid down on a sacrificial foil in place of the glass in the process in the following pargraph. The sacrificial foil is removed after the cell is mounted to a permanent substrate.

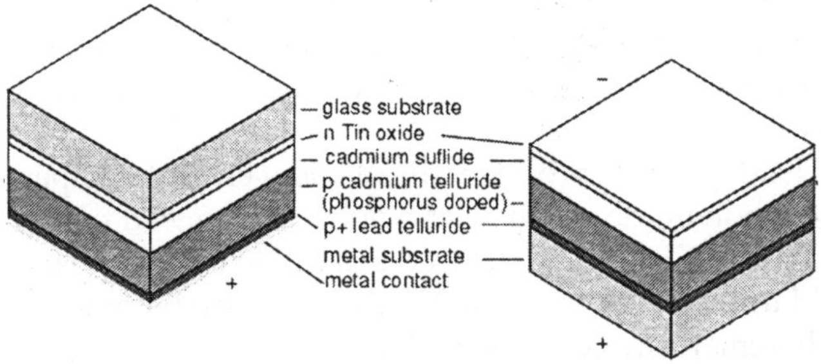

Fig. Cadmium Telluride Solar Cell on Glass or Metal.

A process for depositing cadmium telluride on glass begins with the deposition of N-type transparent, electrically conducive, tin oxide on a glass substrate. The next layer is P-type cadmium telluride; though, N-type or intrinsic may be used. These two layers constitute the NP junction. A P^+ (heavy P-type) layer of lead telluride aids in establishing a low resistance contact. A metal layer makes the final contact to the lead telluride. These layers may be laid down by vacuum deposition, chemical vapor deposition (CVD), screen printing, electrodeposition, or atmospheric pressure chemical vapor deposition (APCVD) in helium. A variation of cadmium telluride is mercury cadmium telluride. Having lower bulk resistance and lower contact resistance improves efficiency over cadmium telluride.

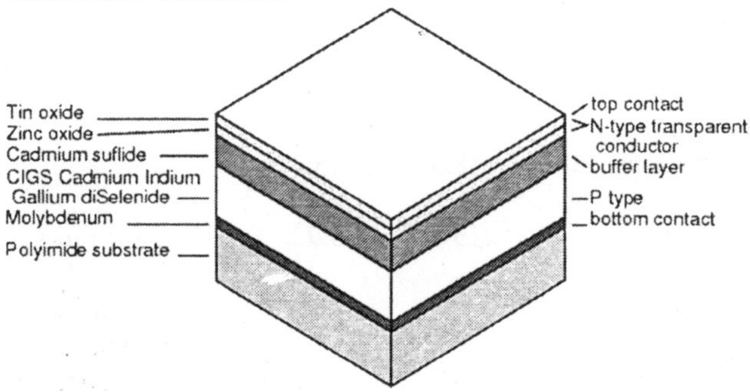

Fig. Cadmium Indium Gallium diSelenide solar cell (CIGS)

Cadmium Indium Gallium diSelenide:

A most promising thin film solar cell at this time (2007) is manufactured on a ten inch wide roll of flexible polyimide–

Cadmium Indium Gallium diSelenide (CIGS). It has a spectacular efficiency of 10%. Though, commercial grade crystalline silicon cells surpassed this decades ago, CIGS should be cost competitive.

The deposition processes are at a low enough temperature to use a polyimide polymer as a substrate instead of metal or glass. The CIGS is manufactured in a roll to roll process, which should drive down costs. GIGS cells may also be produced by an inherently low cost electrochemical process.

Solar Cell Properties

Solar cell type	*Maximum efficiency*	*Practical efficiency*	*Notes*
Selenium, polycrystalline	0.7%	—	1883, Charles Fritts
Silicon, single crystal	—	4%	1950's, first silicon solar cell
Silicon, single crystal PERL, terrestrial, space	25%—		solar cars, cost=100x commercial
Silicon, single crystal, commercial terrestrial	24%	14-17%	$5 $10/peak watt
Cypress Semiconductor, Sunpower, silicon single crystal	21.5%	19%	all contacts on cell back
Gallium Indium Phosphide/ Gallium Arsenide/Germanium, single crystal, multilayer	—	32%	Preferred for space.
Advanced terrestrial version of above.	—	40.7%	Uses optical concentrator.
Silicon, multi crystalline	18.5%	15.5%	—
Thin films,	–	–	–
Silicon, amorphous	13%	5-7%	Degrades in sun light. Good indoors for calculators or cloudy outdoors.
Cadmium telluride, polycrystalline	16%	—	glass or metal substrate
Copper indium arsenide diselenide, polycrystalline	18%	10%	10 inch flexible polymer web.
Organic polymer, 100% plastic	4.5%	—	R&D project

Varicap or Varactor Diodes

A variable capacitance diode is known as a *varicap diode* or as a *varactor*. If a diode is reverse biased, an insulating depletion region forms between the two semiconductive layers. In many diodes the width of the depletion region may be changed by varying the reverse bias. This varies the capacitance. This effect is accentuated in varicap diodes. The schematic symbols is shown in Figure below, one of which is packaged as common cathode dual diode.

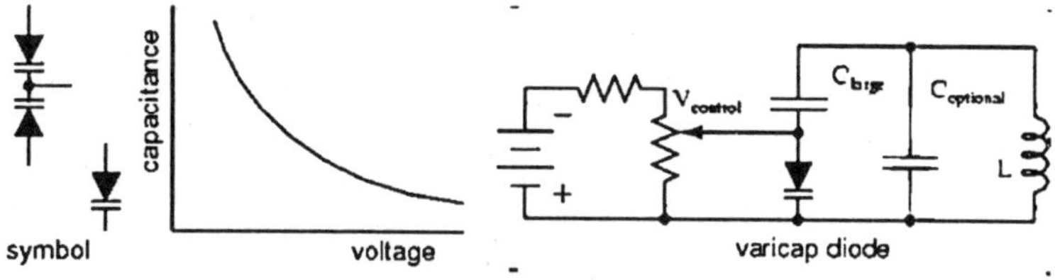

Fig. Varicap diode: Capacitance Varies with Reverse Bias. This Varies the Frequency of a Resonant Network.

If a varicap diode is part of a resonant circuit as in Figure above, the frequency my be varied with a control voltage, $V_{control}$.

A large capacitance, low X_c, in series with the varicap prevents $V_{control}$ from being shorted out by inductor L. As long as the series capacitor is large, it has minimal effect on the frequency of resonant circuit. $C_{optional}$ may be used to set the center resonant frequency. $V_{control}$ can then vary the frequency about this point. Note that the required active circuitry to make the resonant network oscillate is not shown. For an example of a varicap diode tuned AM radio receiver see "electronic varicap diode tuning," Ch 9

Some varicap diodes may be referred to as abrupt, hyperabrupt, or super hyper abrupt. These refer to the change in junction capacitance with changing reverse bias as being abrupt or hyper-abrupt, or super hyperabrupt. These diodes offer a relatively large change in capacitance. This is useful when oscillators or filters are swept over a large frequency range. Varying the bias of abrupt varicaps over the rated limits, changes capacitance by a 4:1 ratio, hyperabrupt by 10:1, super hyperabrupt by 20:1.

Varactor diodes may be used in frequency multiplier circuits. See "Practical analog semiconductor circuits," Varactor multiplier

Snap Diode

The *snap diode,* also known as the *step recovery diode* is designed for use in high ratio frequency multipliers up to 20 gHz. When the diode is forward biased, charge is stored in the PN junction. This charge is drawn out as the diode is reverse biased. The diode looks like a low impedance current source during forward bias. When reverse bias is applied it still looks like a low impedance source until all the charge is withdrawn. It then "snaps" to a high impedance state causing a voltage impulse, rich in harmonics. An applications is a comb generator, a generator of many harmonics. Moderate power 2x and 4x multipliers are another application.

PIN Diodes

A *PIN diode* is a fast low capacitance switching diode. Do not confuse a PIN switching diode with a PIN photo diode here. A PIN diode is manufactured like a silicon switching diode with an intrinsic region added between the PN junction layers. This yields a thicker depletion region, the insulating layer at the junction of a reverse biased diode. This results in lower capacitance than a reverse biased switching diode.

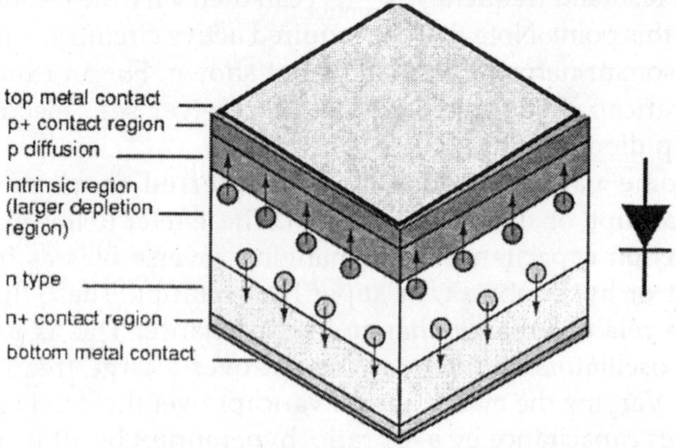

Fig. Pin Diode: Cross Section Aligned with Schematic Symbol.

PIN diodes are used in place of switching diodes in radio frequency (RF) applications, for example, a T/R switch here. The 1n4007 1000 V, 1 A general purpose power diode is reported to be useable as a PIN switching diode. The high voltage rating of this diode is achieved by the inclusion of an intrinsic layer dividing the PN junction. This intrinsic layer makes the 1n4007 a PIN diode. Another PIN diode application is a the antenna switch here for a direction finder receiver.

PIN diodes serve as variable resistors when the forward bias is varied. One such application is the voltage variable attenuator here. The low capacitance characteristic of PIN diodes, extends the frequency flat response of the attenuator to microwave frequencies.

IMPATT Diode

IMPact Avalanche Transit Time diode is a high power radio frequency (RF) generator operating from 3 to 100 gHz. IMPATT diodes are fabricated from silicon, gallium arsenide, or silicon carbide.

An IMPATT diode is reverse biased above the breakdown voltage. The high doping levels produce a thin depletion region. The resulting high electric field rapidly accelerates carriers which free other carriers in collisions with the crystal lattice. Holes are swept into the P_+ region. Electrons drift toward the N regions. The cascading effect creates an avalanche current which increases even as voltage across the junction decreases. The pulses of current lag the voltage peak across the junction. A "negative resistance" effect in conjunction with a resonant circuit produces oscillations at high power levels (high for semiconductors).

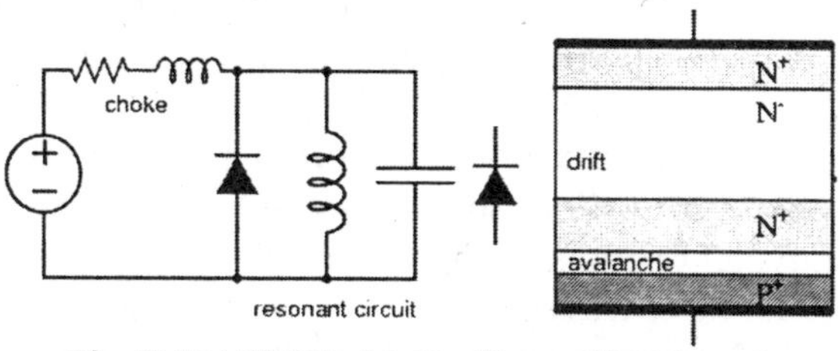

Fig. IMPATT Diode: Oscillator Circuit and Heavily Doped P and N layers.

The resonant circuit in the schematic diagram of Figure above is the lumped circuit equivalent of a waveguide section, where the IMPATT diode is mounted. DC reverse bias is applied through a choke which keeps RF from being lost in the bias supply. This may be a section of waveguide known as a bias Tee. Low power RADAR transmitters may use an IMPATT diode as a power source. They are too noisy for use in the receiver.

Gunn Diode

A *gunn diode* is solely composed of N-type semiconductor. As such, it is not a true diode. Figure below shows a lightly doped N_- layer surrounded by heavily doped N+ layers. A voltage applied across the N-type gallium arsenide gunn diode creates a strong electric field across the lightly doped N^- layer.

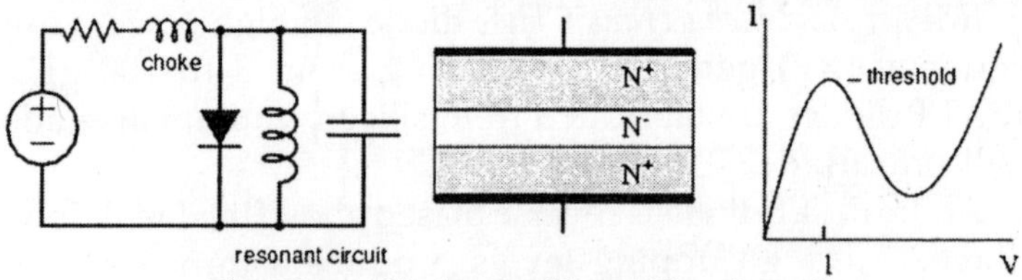

Fig. Gunn diode: Oscillator Circuit and Cross Section of only N-type Semiconductor Diode.

As voltage is increased, conduction increases due to electrons in a low energy conduction band. As voltage is increased beyond the threshold of approximately 1 V, electrons move from the lower conduction band to the higher energy conduction band where they no longer contribute to conduction. In other words, as voltage increases, current decreases, a negative resistance condition. The oscillation frequency is determined by the transit time of the conduction electrons, which is inversely related to the thickness of the N^- layer.

The frequency may be controlled to some extent by embedding the gunn diode into a resonant circuit. The lumped circuit equivalent shown in Figure above is actually a coaxial transmission line or waveguide. Gallium arsenide gunn diodes are available for operation from 10 to 200 gHz at 5 to 65 mw power. Gunn diodes may also serve as amplifiers.

Shockley Diode

The *Shockley diode*is a 4-layer thyristor used to trigger larger thyristors. It only conducts in one direction when triggered by a voltage exceeding the *breakover voltage,* about 20 V. The Shockley Diode. The bidirectional version is called a *diac.*

Constant-Current Diodes

A *constant-current diode,* also known as a *current-limiting diode,* or *current-regulating diode,* does exactly what its name implies: it regulates current through it to some maximum level. The constant current diode is a two terminal version of a JFET. If we try to force more current through a constant-current diode than its current-regulation point, it simply "fights back" by dropping more voltage. If we were to build the circuit in Figure below(a) and plot diode current against diode voltage, we'd get a graph that rises at first and then levels off at the current regulation point as in Figure below(b).

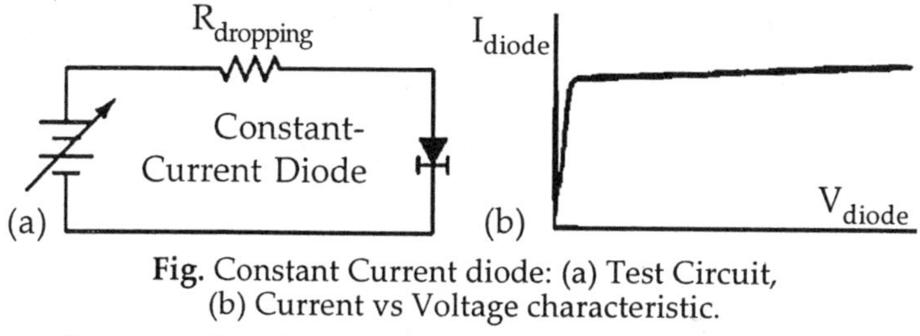

Fig. Constant Current diode: (a) Test Circuit, (b) Current vs Voltage characteristic.

One application for a constant-current diode is to automatically limit current through an LED or laser diode over a wide range of power supply voltages as in Figure below.

Of course, the constant-current diode's regulation point should be chosen to match the LED or laser diode's optimum forward current. This is especially important for the laser diode, not so much for the LED, as regular LEDs tend to be more tolerant of forward current variations.

Another application is in the charging of small secondary-cell batteries, where a constant charging current leads to predictable charging times. Of course, large secondary-cell battery banks might also benefit from constant-current charging, but constant-current diodes tend to be very small devices, limited to regulating currents in the milliamp range.

OTHER DIODE TECHNOLOGIES

SiC Diodes

Diodes manufactured from silicon carbide are capable of high temperature operation to 400°C. This could be in a high temperature environment: down hole oil well logging, gas turbine engines, auto engines. Or, operation in a moderate environment at high power dissipation. Nuclear and space applications are promising as SiC is 100 times more resistant to radiation compared with silicon. SiC is a better conductor of heat than any metal. Thus, SiC is better than silicon at conducting away heat. Breakdown voltage is several kV. SiC power devices are expected to reduce electrical energy losses in the power industry by a factor of 100.

Polymer Diode

Diodes based on organic chemicals have been produce using low temperature processes. Hole rich and electron rich conductive polymers may be ink jet printed in layers. Most of the research and development is of the *organic LED* (OLED). However, development of inexpensive printable organic RFID (radio frequency identification) tags is on going. In this effort, a pentacene organic rectifier has been operated at 50 MHz. Rectification to 800 MHz is a development goal. An inexpensive *metal insulator metal* (MIM) diode acting like a back-to-back zener diode clipper has been delveloped. Also, a tunnel diode like device has been fabricated.

Chapter 3

Bipolar Junction Transistors

The invention of the bipolar transistor in 1948 ushered in a revolution in electronics. Technical feats previously requiring relatively large, mechanically fragile, power-hungry vacuum tubes were suddenly achievable with tiny, mechanically rugged, power-thrifty specks of crystalline silicon. This revolution made possible the design and manufacture of lightweight, inexpensive electronic devices that we now take for granted. Understanding how transistors function is of paramount importance to anyone interested in understanding modern electronics.

A bipolar transistor consists of a three-layer "sandwich" of doped (extrinsic) semiconductor materials, either P-N-P or N-P-N.

Each layer forming the transistor has a specific name, and each layer is provided with a wire contact for connection to a circuit. Shown here are schematic symbols and physical diagrams of these two transistor types:

PNP **transistor**

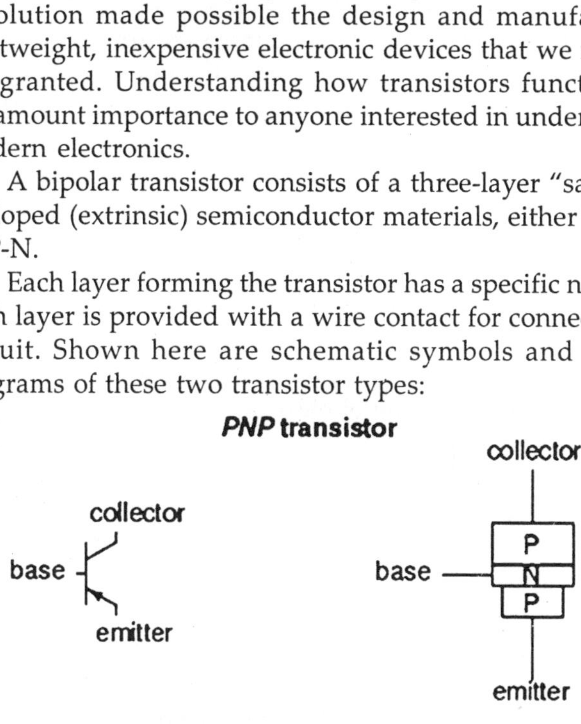

schematic symbol　　　　*physical diagram*

***NPN* transistor**

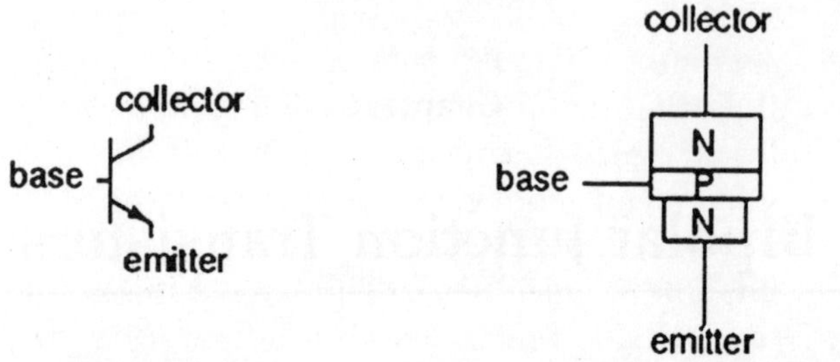

schematic symbol *physical diagram*

The only functional difference between a PNP transistor and an NPN transistor is the proper biasing (polarity) of the junctions when operating. For any given state of operation, the current directions and voltage polarities for each type of transistor are exactly opposite each other.

Bipolar transistors work as current-controlled current *regulators*. In other words, they restrict the amount of current that can go through them according to a smaller, controlling current.

The main current that is *controlled* goes from collector to emitter, or from emitter to collector, depending on the type of transistor it is (PNP or NPN, respectively). The small current that *controls* the main current goes from base to emitter, or from emitter to base, once again depending on the type of transistor it is (PNP or NPN, respectively). According to the confusing standards of semiconductor symbology, the arrow always points *against* the direction of electron flow:

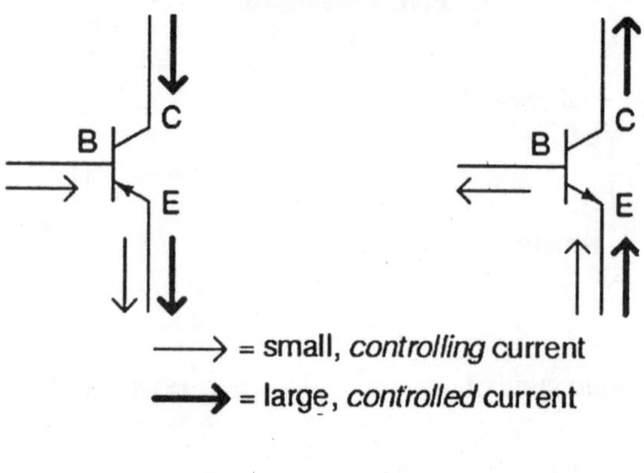

Bipolar transistors are called *bi*polar because the main flow of electrons through them takes place in *two* types of semiconductor material: P and N, as the main current goes from emitter to collector (or vice versa). In other words, two types of charge carriers — electrons and holes — comprise this main current through the transistor.

As you can see, the *controlling* current and the *controlled* current always mesh together through the emitter wire, and their electrons always flow *against* the direction of the transistor's arrow. This is the first and foremost rule in the use of transistors:

All currents must be going in the proper directions for the device to work as a current regulator. The small, controlling current is usually referred to simply as the *base current* because it is the only current that goes through the base wire of the transistor. Conversely, the large, controlled current is referred to as the *collector current* because it is the only current that goes through the collector wire. The emitter current is the sum of the base and collector currents, in compliance with Kirchhoff's Current Law.

If there is no current through the base of the transistor, it shuts off like an open switch and prevents current through the collector. If there is a base current, then the transistor turns on like a closed switch and allows a proportional amount of current through the collector. Collector current is primarily limited by the base current, regardless of the amount of voltage available to push it. The next section will explore in more detail the use of bipolar transistors as switching elements.

THE TRANSISTOR AS A SWITCH

Because a transistor's collector current is proportionally limited by its base current, it can be used as a sort of current-controlled switch. A relatively small flow of electrons sent through the base of the transistor has the ability to exert control over a much larger flow of electrons through the collector.

Suppose we had a lamp that we wanted to turn on and off by means of a switch. Such a circuit would be extremely simple:

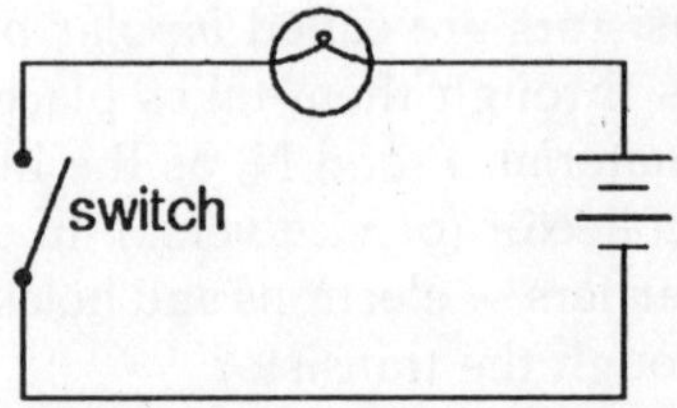

For the sake of illustration, let's insert a transistor in place of the switch to show how it can control the flow of electrons through the lamp. Remember that the controlled current through a transistor must go between collector and emitter. Since its the current through the lamp that we want to control, we must position the collector and emitter of our transistor where the two contacts of the switch are now. We must also make sure that the lamp's current will move *against* the direction of the emitter arrow symbol to ensure that the transistor's junction bias will be correct:

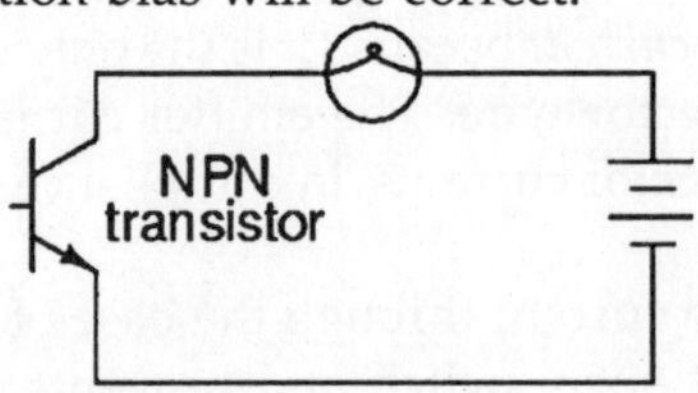

In this example I happened to choose an NPN transistor. A PNP transistor could also have been chosen for the job, and its application would look like this:

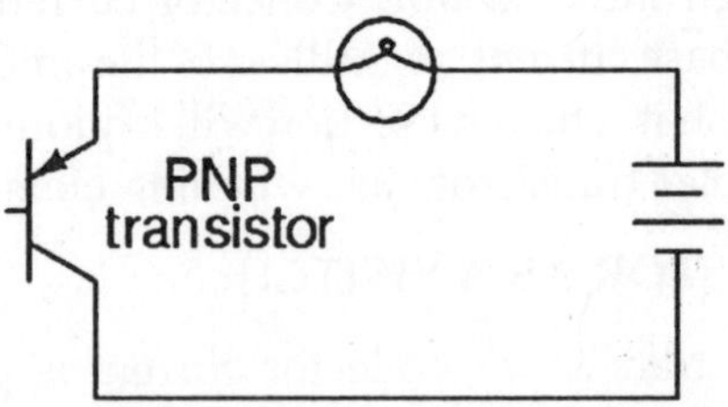

The choice between NPN and PNP is really arbitrary. All that matters is that the proper current directions are maintained for the sake of correct junction biasing (electron flow going *against* the transistor symbol's arrow).

Going back to the NPN transistor in our example circuit, we are faced with the need to add something more so that we can have base current. Without a connection to the base wire of the transistor, base current will be zero, and the transistor cannot turn on, resulting in a lamp that is always off. Remember that for an NPN transistor, base current must consist of electrons flowing from emitter to base (against the emitter arrow symbol, just like the lamp current). Perhaps the simplest thing to do would be to connect a switch between the base and collector wires of the transistor like this:

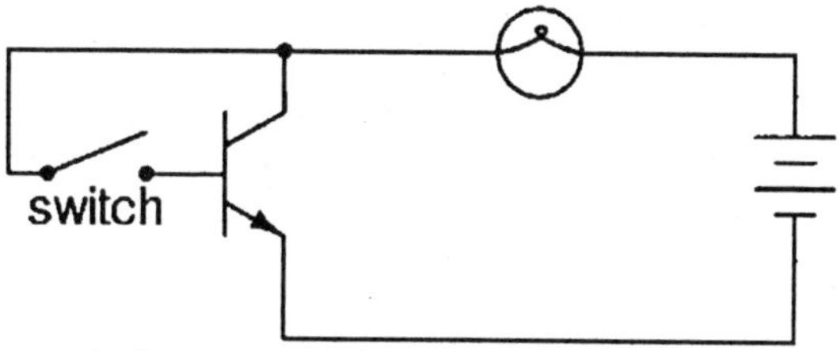

If the switch is open, the base wire of the transistor will be left "floating" (not connected to anything) and there will be no current through it. In this state, the transistor is said to be *cutoff*. If the switch is closed, however, electrons will be able to flow from the emitter through to the base of the transistor, through the switch and up to the left side of the lamp, back to the positive side of the battery. This base current will enable a much larger flow of electrons from the emitter through to the collector, thus lighting up the lamp. In this state of maximum circuit current, the transistor is said to be *saturated*.

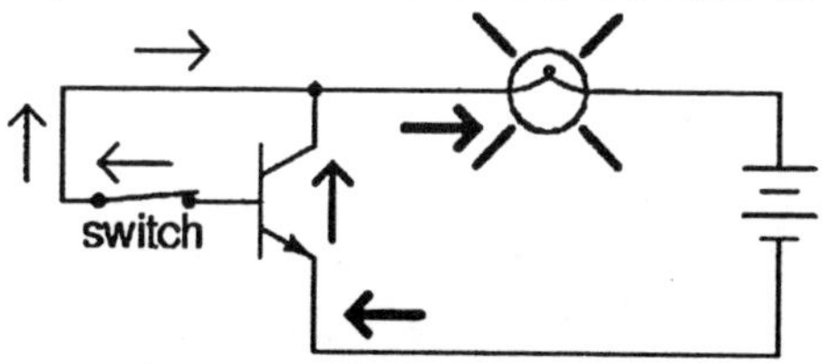

Of course, it may seem pointless to use a transistor in this capacity to control the lamp. After all, we're still using a switch in the circuit, aren't we? If we're still using a switch to control the lamp — if only indirectly — then what's the point of having a transistor to control the current? Why not just go back to our original circuit and use the switch directly to control the lamp current?

There are a couple of points to be made here, actually. First is the fact that when used in this manner, the switch contacts need only handle what little base current is necessary to turn the transistor on, while the transistor itself handles the majority of the lamp's current. This may be an important advantage if the switch has a low current rating: a small switch may be used to control a relatively high-current load. Perhaps more importantly, though, is the fact that the current-controlling behavior of the transistor enables us to use something completely different to turn the lamp on or off. Consider this example, where a solar cell is used to control the transistor, which in turn controls the lamp:

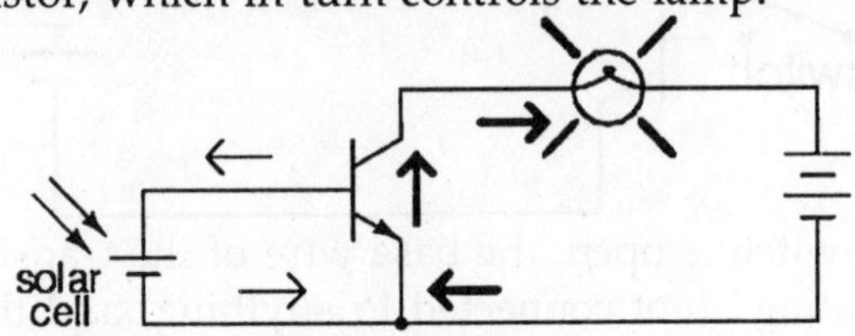

Or, we could use a thermocouple to provide the necessary base current to turn the transistor on:

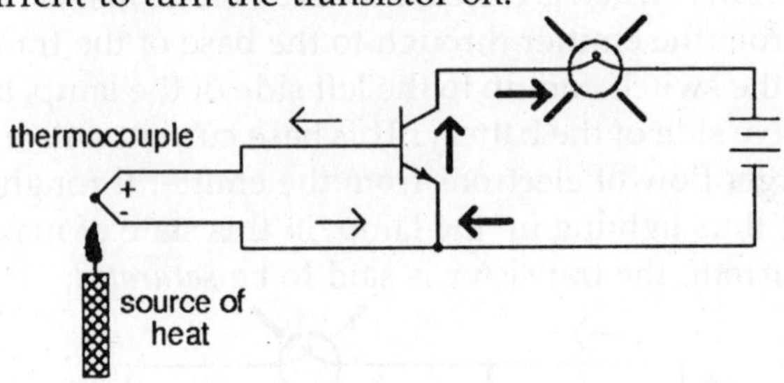

Even a microphone of sufficient voltage and current output could be used to turn the transistor on, provided its output is rectified from AC to DC so that the emitter-base PN junction within the transistor will always be forward-biased:

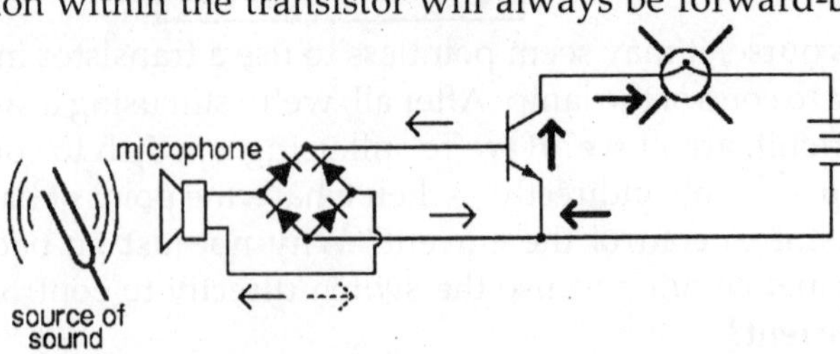

The point should be quite apparent by now: *any* sufficient source of DC current may be used to turn the transistor on, and that source of current need only be a fraction of the amount of current needed to energize the lamp.

Here we see the transistor functioning not only as a switch, but as a true *amplifier*: using a relatively low-power signal to *control* a relatively large amount of power. Please note that the actual power for lighting up the lamp comes from the battery to the right of the schematic. It is not as though the small signal current from the solar cell, thermocouple, or microphone is being magically transformed into a greater amount of power. Rather, those small power sources are simply *controlling* the battery's power to light up the lamp.

METER CHECK OF A TRANSISTOR

Bipolar transistors are constructed of a three-layer semiconductor "sandwich," either PNP or NPN. As such, they register as two diodes connected back-to-back when tested with a multimeter's "resistance" or "diode check" functions:

Meter readings will be exactly opposite, of course, for an NPN transistor, with both PN junctions facing the other way. If a multimeter with a "diode check" function is used in this test, it will be found that the emitter-base junction possesses a slightly greater forward voltage drop than the collector-base junction. This forward voltage difference is due to the disparity in doping concentration between the emitter and collector regions of the transistor: the emitter is a much more heavily doped piece of semiconductor material than the collector, causing its junction with the base to produce a higher forward voltage drop.

Knowing this, it becomes possible to determine which wire is which on an unmarked transistor. This is important because transistor packaging, unfortunately, is not standardized. All bipolar transistors have three wires, of course, but the positions of the three wires on the actual physical package are not arranged in any universal, standardized order.

Suppose a technician finds a bipolar transistor and proceeds to measure continuity with a multimeter set in the "diode check" mode. Measuring between pairs of wires and recording the values displayed by the meter, the technician obtains the following data:

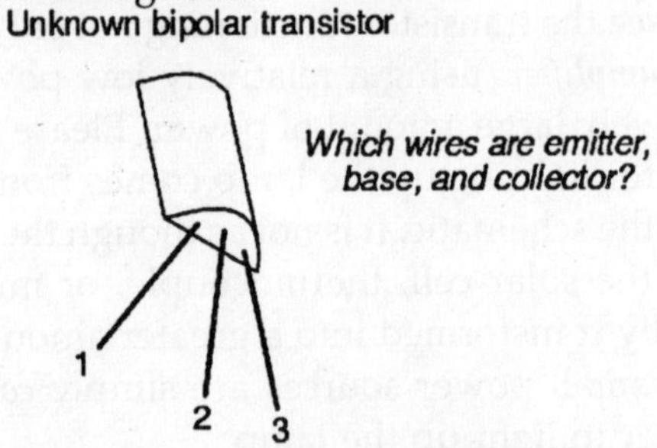

- Meter touching wire 1 (+) and 2 (–): "OL"
- Meter touching wire 1 (–) and 2 (+): "OL"
- Meter touching wire 1 (+) and 3 (–): 0.655 volts
- Meter touching wire 1 (–) and 3 (+): "OL"
- Meter touching wire 2 (+) and 3 (–): 0.621 volts
- Meter touching wire 2 (–) and 3 (+): "OL"

The only combinations of test points giving conducting meter readings are wires 1 and 3 (red test lead on 1 and black test lead on 3), and wires 2 and 3 (red test lead on 2 and black test lead on 3). These two readings *must* indicate forward biasing of the emitter-to-base junction (0.655 volts) and the collector-to-base junction (0.621 volts).

Now we look for the one wire common to both sets of conductive readings. It must be the base connection of the transistor, because the base is the only layer of the three-layer device common to both sets of PN junctions (emitter-base and collector-base).

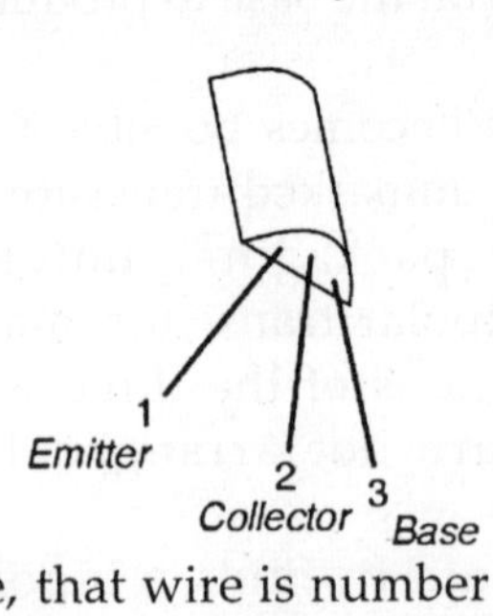

In this example, that wire is number 3, being common to both the 1-3 and the 2-3 test point combinations. In both those sets of meter readings, the *black* (–) meter test lead was touching wire 3, which tells us that the base of this transistor is made of N-type semiconductor material (black = negative). Thus, the transistor is an PNP type with base on wire 3, emitter on wire 1 and collector on wire 2:

Please note that the base wire in this example is *not* the middle lead of the transistor, as one might expect from the three-layer "sandwich" model of a bipolar transistor. This is quite often the case, and tends to confuse new students of electronics. The only way to be sure which lead is which is by a meter check, or by referencing the manufacturer's "data sheet" documentation on that particular part number of transistor.

Knowing that a bipolar transistor behaves as two back-to-back diodes when tested with a conductivity meter is helpful for identifying an unknown transistor purely by meter readings. It is also helpful for a quick functional check of the transistor. If the technician were to measure continuity in any more than two or any less than two of the six test lead combinations, he or she would immediately know that the transistor was defective (or else that it *wasn't* a bipolar transistor but rather something else — a distinct possibility if no part numbers can be referenced for sure identification!). However, the "two diode" model of the transistor fails to explain how or why it acts as an amplifying device.

To better illustrate this paradox, let's examine one of the transistor switch circuits using the physical diagram rather than the schematic symbol to represent the transistor. This way the two PN junctions will be easier to see:

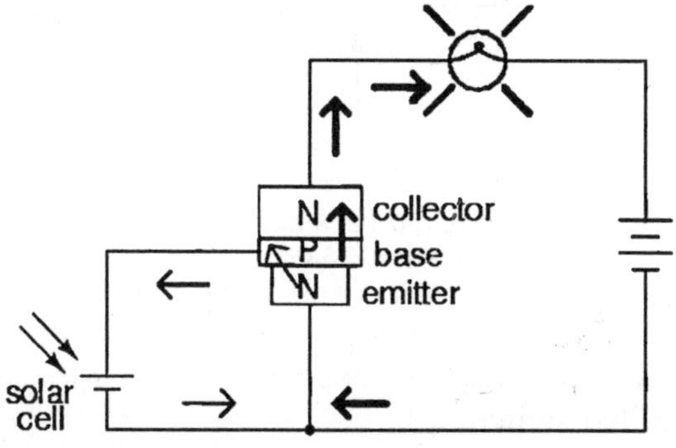

A grey-colored diagonal arrow shows the direction of electron flow through the emitter-base junction. This part makes sense, since the electrons are flowing from the N-type emitter to the P-type base: the junction is obviously forward-biased. However, the base-collector junction is another matter entirely. Notice how the grey-colored thick arrow is pointing in the direction of electron flow (upwards) from base to collector.

With the base made of P-type material and the collector of N-type material, this direction of electron flow is clearly backwards to the direction normally associated with a PN junction! A normal PN junction wouldn't permit this "backward" direction of flow, at least not without offering significant opposition. However, when the transistor is saturated, there is very little opposition to electrons all the way from emitter to collector, as evidenced by the lamp's illumination!

Clearly then, something is going on here that defies the simple "two-diode" explanatory model of the bipolar transistor. When I was first learning about transistor operation, I tried to construct my own transistor from two back-to-back diodes, like this:

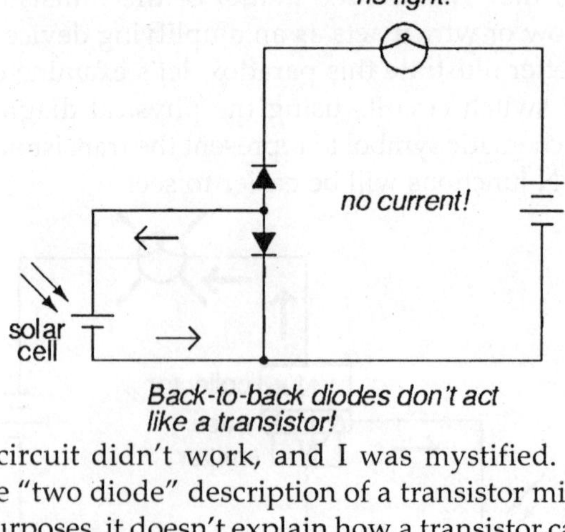

Back-to-back diodes don't act like a transistor!

My circuit didn't work, and I was mystified. However useful the "two diode" description of a transistor might be for testing purposes, it doesn't explain how a transistor can behave as a controlled switch.

What happens in a transistor is this: the reverse bias of the base-collector junction prevents collector current when the transistor is in cutoff mode (that is, when there is no base current). However, when the base-emitter junction is forward biased by the controlling signal, the normally-blocking action of the base-collector junction is overridden and current is permitted through the collector, despite the fact that electrons are going the "wrong way" through that PN junction.

This action is dependent on the quantum physics of semiconductor junctions, and can only take place when the two junctions are properly spaced and the doping concentrations of the three layers are properly proportioned. Two diodes wired in series fail to meet these criteria, and so the top diode can never "turn on" when it is reversed biased, no matter how much current goes through the bottom diode in the base wire loop.

That doping concentrations play a crucial part in the special abilities of the transistor is further evidenced by the fact that collector and emitter are not interchangeable. If the transistor is merely viewed as two back-to-back PN junctions, or merely as a plain N-P-N or P-N-P sandwich of materials, it may seem as though either end of the transistor could serve as collector or emitter.

This, however, is not true. If connected "backwards" in a circuit, a base-collector current will fail to control current between collector and emitter. Despite the fact that both the emitter and collector layers of a bipolar transistor are of the same doping *type* (either N or P), they are definitely not identical!

So, current through the emitter-base junction allows current through the reverse-biased base-collector junction. The action of base current can be thought of as "opening a gate" for current through the collector. More specifically, any given amount of emitter-to-base current *permits a limited amount* of base-to-collector current. For every electron that passes through the emitter-base junction and on through the base wire, there is allowed a certain, restricted number of electrons to pass through the base-collector junction and no more.

In the next section, this current-limiting behavior of the transistor will be investigated in more detail.

ACTIVE MODE OPERATION

When a transistor is in the fully-off state (like an open switch), it is said to be *cutoff*. Conversely, when it is fully conductive between emitter and collector (passing as much current through the collector as the collector power supply and load will allow), it is said to be *saturated*. These are the two modes of operation explored thus far in using the transistor as a switch.

However, bipolar transistors don't have to be restricted to these two extreme modes of operation. As we learned in the previous section, base current "opens a gate" for a limited amount of current through the collector. If this limit for the controlled current is greater than zero but less than the maximum allowed by the power supply and load circuit, the transistor will "throttle" the collector current in a mode somewhere between cutoff and saturation. This mode of operation is called the *active* mode.

An automotive analogy for transistor operation is as follows: *cutoff* is the condition where there is no motive force generated by the mechanical parts of the car to make it move. In cutoff mode, the brake is engaged (zero base current), preventing motion (collector current).

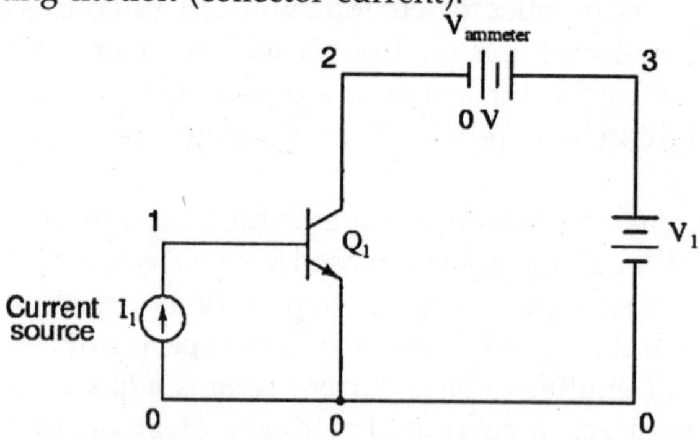

Active mode is when the automobile is cruising at a constant, controlled speed (constant, controlled collector current) as dictated by the driver. *Saturation* is when the automobile is driving up a steep hill that prevents it from going as fast as the driver would wish. In other words, a "saturated" automobile is one where the accelerator pedal is pushed all the way down (base current calling for more collector current than can be provided by the power supply/load circuit).

"Q" is the standard letter designation for a transistor in a schematic diagram, just as "R" is for resistor and "C" is for capacitor. In this circuit, we have an NPN transistor powered by a battery (V_1) and controlled by current through a *current source* (I_1).

A current source is a device that outputs a specific amount of current, generating as much or as little voltage as necessary across its terminals to ensure that exact amount of current through it.

Current sources are notoriously difficult to find in nature (unlike voltage sources, which by contrast attempt to maintain a constant voltage, outputting as much or as little current in the fulfillment of that task), but can be simulated with a small collection of electronic components. As we are about to see, transistors themselves tend to mimic the constant-current behavior of a current source in their ability to *regulate* current at a fixed value.

This voltage/current relationship is entirely different from what we're used to seeing across a resistor. With a resistor, current increases linearly as the voltage across it increases. Here, with a transistor, current from emitter to collector stays limited at a fixed, maximum value no matter how high the voltage across emitter and collector increases.

Often it is useful to superimpose several collector current/voltage graphs for different base currents on the same graph. A collection of curves like this — one curve plotted for each distinct level of base current — for a particular transistor is called the transistor's *characteristic curves*:

Each curve on the graph reflects the collector current of the transistor, plotted over a range of collector-to-emitter voltages, for a given amount of base current.

Since a transistor tends to act as a current regulator, limiting collector current to a proportion set by the base current, it is useful to express this proportion as a standard transistor performance measure.

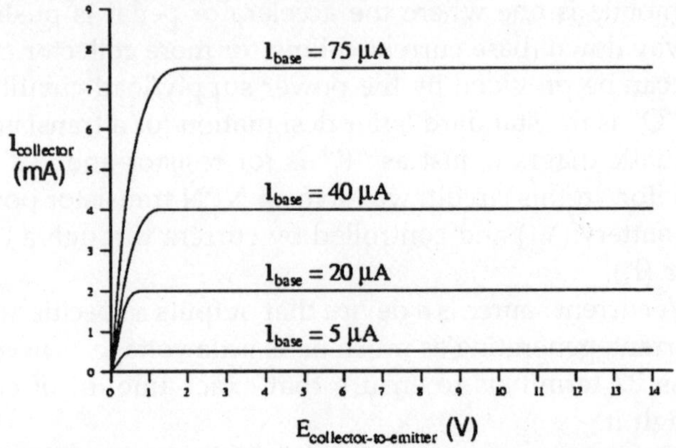

Specifically, the ratio of collector current to base current is know n as the *Beta* ratio (symbolized by the Greek letter β):

$$\beta = \frac{I_{collector}}{I_{base}}.$$

β is also know as h_{fe}: Sometimes the β ratio is designated as "h_{fe}," a label used in a branch of mathematical semiconductor analysis known as "hybrid parameters" which strives to achieve very precise predictions of transistor performance with detailed equations. Hybrid parameter variables are many, but they are all labeled with the general letter "h" and a specific subscript. The variable "h_{fe}" is just another (standardized) way of expressing the ratio of collector current to base current, and is interchangeable with "β." Like all ratios, β is unitless.

β for any transistor is determined by its design: it cannot be altered after manufacture. However, there are so many physical variables impacting β that it is rare to have two transistors of the same design exactly match. If a circuit design relies on equal β ratios between multiple transistors, "matched sets" of transistors may be purchased at extra cost. However, it is generally considered bad design practice to engineer circuits with such dependencies.

It would be nice if the β of a transistor remained stable for all operating conditions, but this is not true in real life. For an actual transistor, the β ratio may vary by a factor of over 3 within its operating current limits. For example, a transistor with advertised β of 50 may actually test with I_c/I_b ratios as low as 30 and as high as 100, depending on the amount of collector current, the transistor's temperature, and frequency of amplified signal, among other factors.

For tutorial purposes it is adequate to assume a constant β for any given transistor, but just realize that real life is not that simple!

Sometimes it is helpful for comprehension to "model" complex electronic components with a collection of simpler, better-understood components. The following is a popular model shown in many introductory electronics texts:

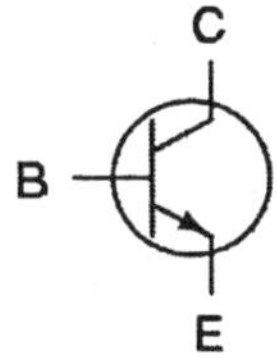

NPN diode-rheostat model

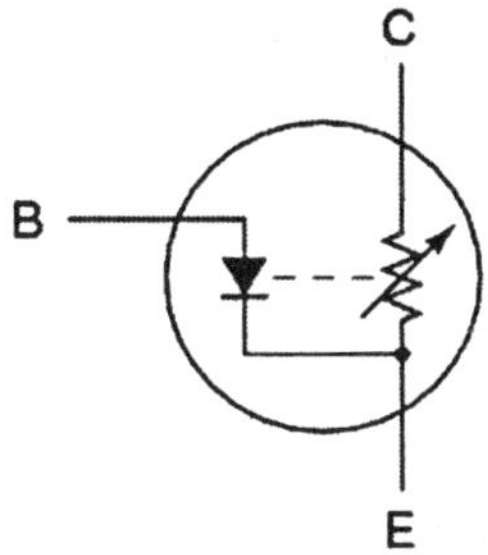

This model casts the transistor as a combination of diode and rheostat (variable resistor). Current through the base-emitter diode controls the resistance of the collector-emitter rheostat (as implied by the dashed line connecting the two components), thus controlling collector current. An NPN transistor is modeled in the figure shown, but a PNP transistor would be only slightly different (only the base-emitter diode would be reversed).

This model succeeds in illustrating the basic concept of transistor amplification: how the base current signal can exert control over the collector current.

However, I personally don't like this model because it tends to mis-communicate the notion of a set amount of collector-emitter resistance for a given amount of base current.

If this were true, the transistor wouldn't *regulate* collector current at all like the characteristic curves show. Instead of the collector current curves flattening out after their brief rise as the collector-emitter voltage increases, the collector current would be directly proportional to collector-emitter voltage, rising steadily in a straight line on the graph.

A better transistor model, is this:

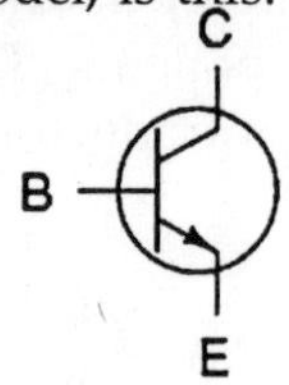

NPN diode-current source model

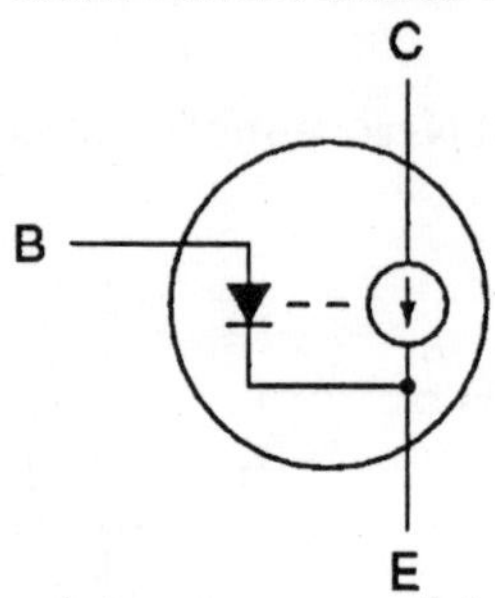

It casts the transistor as a combination of diode and current source, the output of the current source being set at a multiple (β ratio) of the base current. This model is far more accurate in depicting the true input/output characteristics of a transistor: base current establishes a certain amount of collector *current*, rather than a certain amount of collector-emitter *resistance* as the first model implies. Also, this model is favored when performing network analysis on transistor circuits, the current source being a well-understood theoretical component.

Unfortunately, using a current source to model the transistor's current-controlling behavior can be misleading: in no way will the transistor ever act as a *source* of electrical energy, which the current source symbol implies is a possibility.

My own personal suggestion for a transistor model substitutes a constant-current diode for the current source:

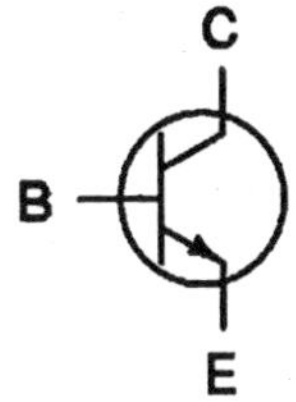

NPN diode-regulating diode model

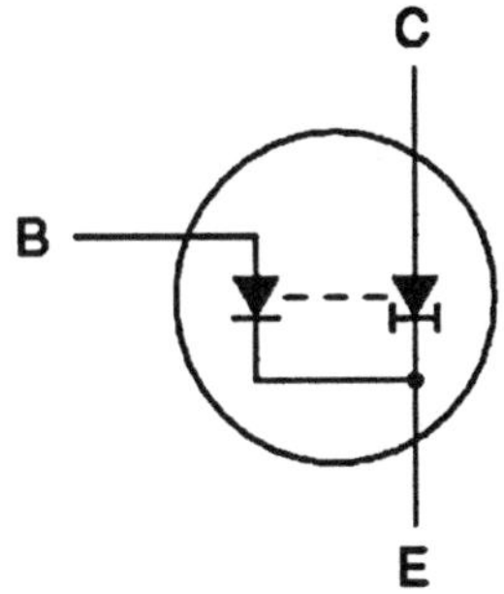

Since no diode ever acts as a *source* of electrical energy, this analogy escapes the false implication of the current source model as a source of power, while depicting the transistor's constant-current behavior better than the rheostat model. Another way to describe the constant-current diode's action would be to refer to it as a *current regulator*, so this transistor illustration of mine might also be described as a *diode-current regulator* model.

The greatest disadvantage I see to this model is the relative obscurity of constant-current diodes. Many people may be unfamiliar with their symbology or even of their existence, unlike either rheostats or current sources, which are commonly known.

THE COMMON-EMITTER AMPLIFIER

At the beginning we saw how transistors could be used as switches, operating in either their "saturation" or "cutoff" modes. In the last section we saw how transistors behave within their "active" modes, between the far limits of saturation and cutoff. Because transistors are able to control current in an analog (infinitely divisible) fashion, they find use as amplifiers for analog signals.

One of the simpler transistor amplifier circuits to study is the one used previously for illustrating the transistor's switching ability:

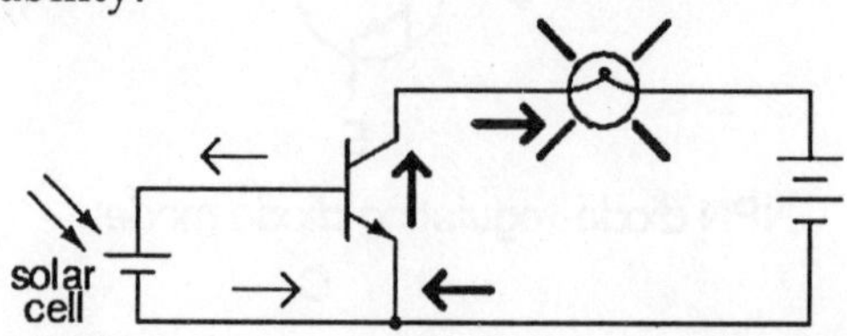

It is called the *common-emitter* configuration because (ignoring the power supply battery) both the signal source and the load share the emitter lead as a common connection point.

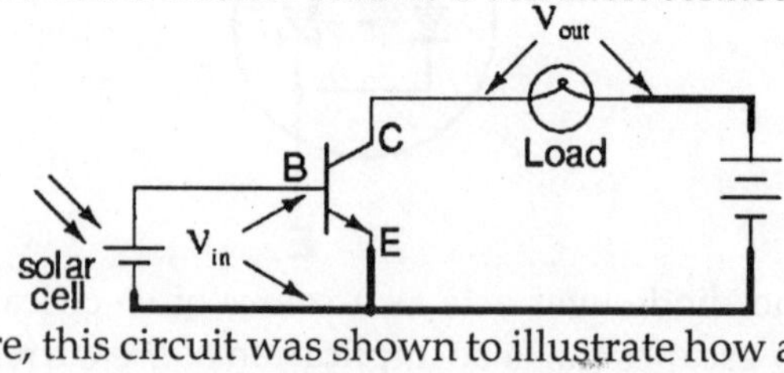

Before, this circuit was shown to illustrate how a relatively small current from a solar cell could be used to saturate a transistor, resulting in the illumination of a lamp. Knowing now that transistors are able to "throttle" their collector currents according to the amount of base current supplied by an input signal source, we should be able to see that the brightness of the lamp in this circuit is controllable by the solar cell's light exposure.

When there is just a little light shone on the solar cell, the lamp will glow dimly. The lamp's brightness will steadily increase as more light falls on the solar cell.

Suppose that we were interested in using the solar cell as a light intensity instrument. We want to be able to measure the intensity of incident light with the solar cell by using its output current to drive a meter movement. It is possible to directly connect a meter movement to a solar cell for this purpose. In fact, the simplest light-exposure meters for photography work are designed like this:

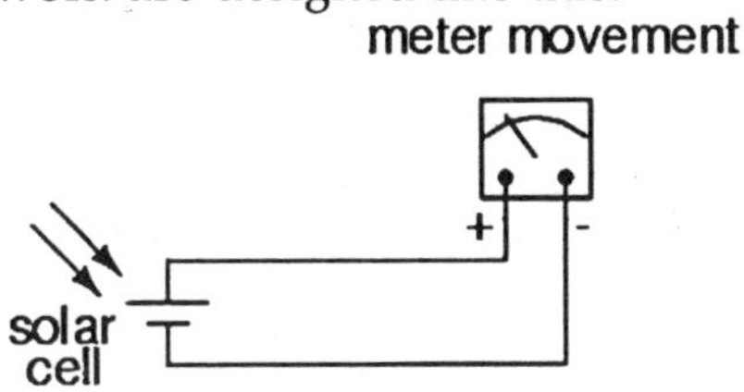

While this approach might work for moderate light intensity measurements, it would not work as well for low light intensity measurements. Because the solar cell has to supply the meter movement's power needs, the system is necessarily limited in its sensitivity. Supposing that our need here is to measure very low-level light intensities, we are pressed to find another solution.

Perhaps the most direct solution to this measurement problem is to use a transistor to *amplify* the solar cell's current so that more meter movement needle deflection may be obtained for less incident light. Consider this approach:

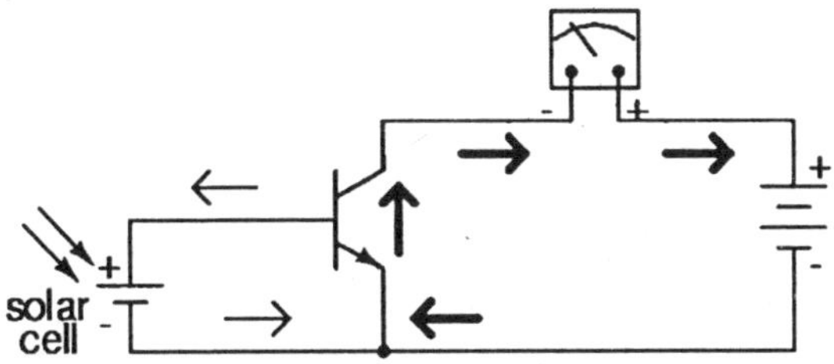

Current through the meter movement in this circuit will be β times the solar cell current. With a transistor β of 100, this represents a substantial increase in measurement sensitivity. It is prudent to point out that the additional power to move the meter needle comes from the battery on the far right of the circuit, not the solar cell itself. All the solar cell's current does is *control* battery current to the meter to provide a greater meter reading than the solar cell could provide unaided.

Because the transistor is a current-regulating device, and because meter movement indications are based on the amount of current through their movement coils, meter indication in this circuit should depend only on the amount of current from the solar cell, not on the amount of voltage provided by the battery. This means the accuracy of the circuit will be independent of battery condition, a significant feature! All that is required of the battery is a certain minimum voltage and current output ability to be able to drive the meter full-scale if needed.

Another way in which the common-emitter configuration may be used is to produce an output *voltage* derived from the input signal, rather than a specific output *current*. Let's replace the meter movement with a plain resistor and measure voltage between collector and emitter:

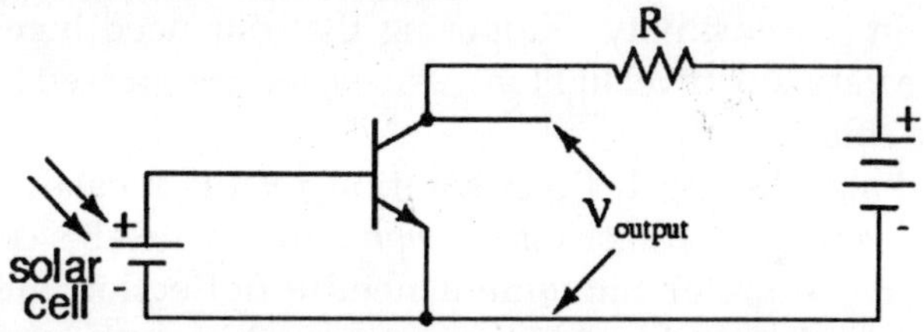

With the solar cell darkened (no current), the transistor will be in cutoff mode and behave as an open switch between collector and emitter. This will produce maximum voltage drop between collector and emitter for maximum V_{output}, equal to the full voltage of the battery.

At full power (maximum light exposure), the solar cell will drive the transistor into saturation mode, making it behave like a closed switch between collector and emitter. The result will be minimum voltage drop between collector and emitter, or almost zero output voltage. In actuality, a saturated transistor can never achieve zero voltage drop between collector and emitter due to the two PN junctions through which collector current must travel. However, this "collector-emitter saturation voltage" will be fairly low, around several tenths of a volt, depending on the specific transistor used.

For light exposure levels somewhere between zero and maximum solar cell output, the transistor will be in its active mode, and the output voltage will be somewhere between zero and full battery voltage. An important quality to note here about the common-emitter configuration is that the output voltage is *inversely proportional* to the input signal strength. That is, the output voltage decreases as the input signal increases. For this reason, the common-emitter amplifier configuration is referred to as an *inverting* amplifier.

The simulation plots both the input voltage (an AC signal of 1.5 volt peak amplitude and 2000 Hz frequency) and the current through the 15 volt battery, which is the same as the current through the speaker. What we see here is a full AC sine wave alternating in both positive and negative directions, and a half-wave output current waveform that only pulses in one direction. If we were actually driving a speaker with this waveform, the sound produced would be horribly distorted.

What's wrong with the circuit? Why won't it faithfully reproduce the entire AC waveform from the microphone? The answer to this question is found by close inspection of the transistor diode-regulating diode model:

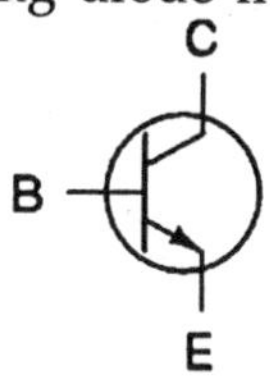

NPN diode-regulating diode model

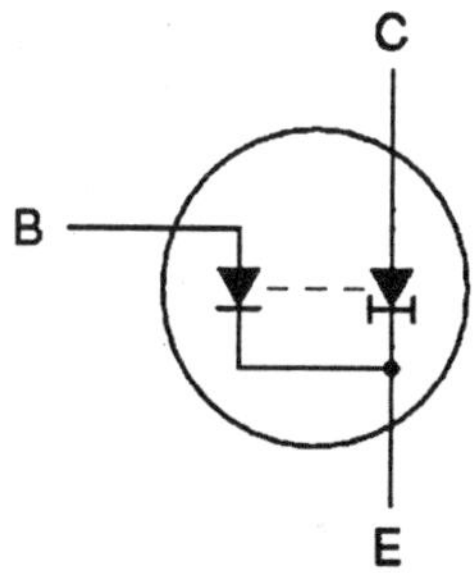

Collector current is controlled, or regulated, through the constant-current mechanism according to the pace set by the current through the base-emitter diode. Note that both current paths through the transistor are monodirectional: *one way only!* Despite our intent to use the transistor to amplify an *AC* signal, it is essentially a *DC* device, capable of handling currents in a single direction only. We may apply an AC voltage input signal between the base and emitter, but electrons cannot flow in that circuit during the part of the cycle that reverse-biases the base-emitter diode junction.

Therefore, the transistor will remain in cutoff mode throughout that portion of the cycle. It will "turn on" in its active mode only when the input voltage is of the correct polarity to forward-bias the base-emitter diode, and only when that voltage is sufficiently high to overcome the diode's forward voltage drop. Remember that bipolar transistors are *current-controlled devices*: they regulate collector current based on the existence of base-to-emitter *current*, not base-to-emitter *voltage*.

The only way we can get the transistor to reproduce the entire waveform as current through the speaker is to keep the transistor in its active mode the entire time. This means we must maintain current through the base during the entire input waveform cycle. Consequently, the base-emitter diode junction must be kept forward-biased at all times. Fortunately, this can be accomplished with the aid of a DC *bias voltage* added to the input signal. By connecting a sufficient DC voltage in series with the AC signal source, forward-bias can be maintained at all points throughout the wave cycle:

The following illustration is another view of the same circuit, this time with a few oscilloscopes ("scopemeters") connected at crucial points to display all the pertinent signals:

The need for biasing a transistor amplifier circuit to obtain full waveform reproduction is an important consideration.

PNP transistors are just as valid to use as NPN in *any* amplifier configuration, so long as the proper polarity and current directions are maintained, and the common-emitter amplifier is no exception. The inverting behavior and gain properties of a PNP transistor amplifier are the same as its NPN counterpart, just the polarities are different:

THE COMMON-COLLECTOR AMPLIFIER

Our next transistor configuration to study is a bit simpler in terms of gain calculations. Called the *common-collector* configuration, its schematic diagram looks like this:

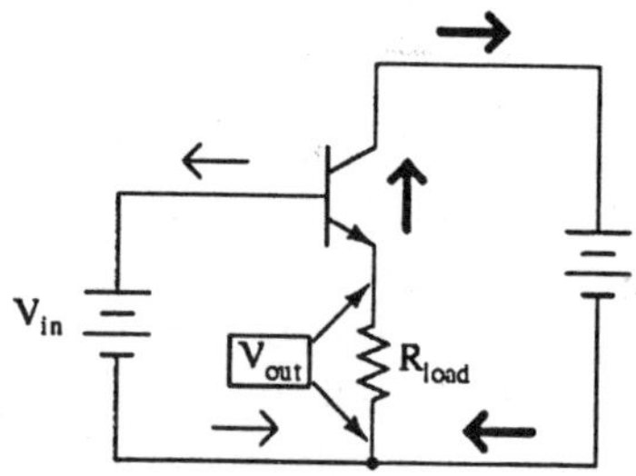

It is called the *common-collector* configuration because (ignoring the power supply battery) both the signal source and the load share the collector lead as a common connection point:

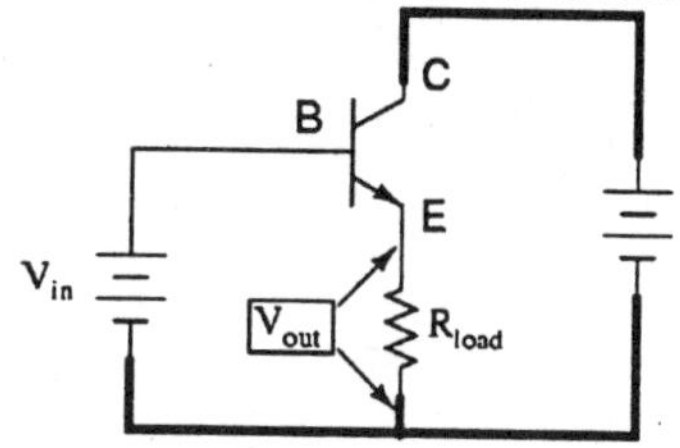

It should be apparent that the load resistor in the common-collector amplifier circuit receives both the base and collector currents, being placed in series with the emitter. Since the emitter lead of a transistor is the one handling the most current (the sum of base and collector currents, since base and collector currents always mesh together to form the emitter current), it would be reasonable to presume that this amplifier will have a very large current gain (maximum output current for minimum input current). This presumption is indeed correct: the current gain for a common-collector amplifier is quite large, larger than any other transistor amplifier configuration. However, this is not necessarily what sets it apart from other amplifier designs.

Here's another view of the circuit, this time with oscilloscopes connected to several points of interest:

Since this amplifier configuration doesn't provide any voltage gain (in fact, in practice it actually has a voltage gain of slightly *less* than 1), its only amplifying factor is current. The common-emitter amplifier configuration examined in the previous section had a current gain equal to the β of the transistor, being that the input current went through the base and the output (load) current went through the collector, and β by definition is the ratio between the collector and base currents. In the common-collector configuration, though, the load is situated in series with the emitter, and thus its current is equal to the emitter current. With the emitter carrying collector current *and* base current, the load in this type of amplifier has all the current of the collector running through it *plus* the input current of the base. This yields a current gain of β plus 1:

$$A_1 = \frac{I_{emitter}}{I_{base}}$$

$$A_1 = \frac{I_{emitter} + I_{base}}{I_{base}}$$

$$A_1 = \frac{I_{emitter}}{I_{base}} + 1$$

$$A_1 = \beta + 1.$$

Once again, PNP transistors are just as valid to use in the common-collector configuration as NPN transistors. The gain calculations are all the same, as is the non-inverting behavior of the amplifier. The only difference is in voltage polarities and current directions:

A popular application of the common-collector amplifier is for regulated DC power supplies, where an unregulated (varying) source of DC voltage is clipped at a specified level to supply regulated (steady) voltage to a load. Of course, zener diodes already provide this function of voltage regulation:

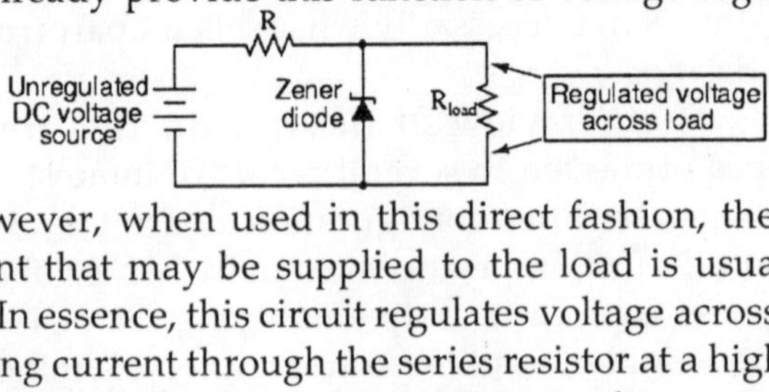

However, when used in this direct fashion, the amount of current that may be supplied to the load is usually quite limited. In essence, this circuit regulates voltage across the load by keeping current through the series resistor at a high enough level to drop all the excess power source voltage across it, the zener diode drawing more or less current as necessary to keep the voltage across itself steady. For high-current loads, an plain zener diode voltage regulator would have to be capable of shunting a lot of current through the diode in order to be effective at regulating load voltage in the event of large load resistance or voltage source changes.

One popular way to increase the current-handling ability of a regulator circuit like this is to use a common-collector transistor to amplify current to the load, so that the zener diode circuit only has to handle the amount of current necessary to drive the base of the transistor:

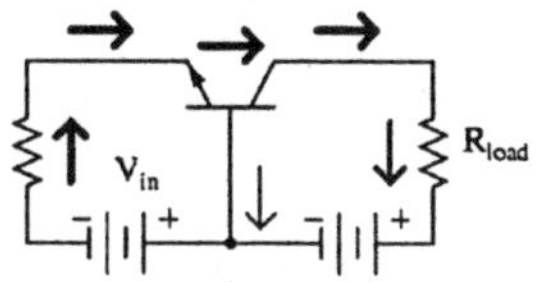

There's really only one caveat to this approach: the load voltage will be approximately 0.7 volts less than the zener diode voltage, due to the transistor's 0.7 volt base-emitter drop. However, since this 0.7 volt difference is fairly constant over a wide range of load currents, a zener diode with a 0.7 volt higher rating can be chosen for the application.

Sometimes the high current gain of a single-transistor, common-collector configuration isn't enough for a particular application. If this is the case, multiple transistors may be staged together in a popular configuration known as a *Darlington pair*, just an extension of the common-collector concept:

An NPN "Darlington pair"

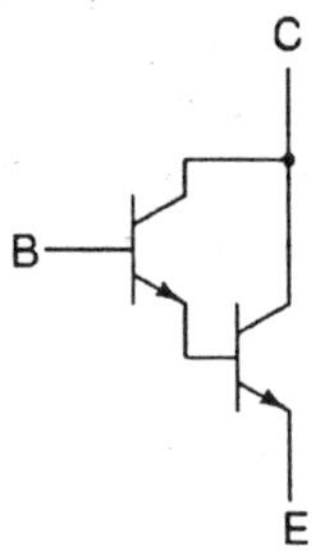

Darlington pairs essentially place one transistor as the common-collector load for another transistor, thus multiplying their individual current gains. Base current through the upper-left transistor is amplified through that transistor's emitter, which is directly connected to the base of the lower-right transistor, where the current is again amplified. The overall current gain is as follows:

Darlington pair current gain

$$A_1 = (\beta_1 + 1)(\beta_2 + 1)$$

Where,

β_1 = Beta of first transistor

β_2 = Beta of second transistor

Voltage gain is still nearly equal to 1 if the entire assembly is connected to a load in common-collector fashion, although the load voltage will be a full 1.4 volts less than the input voltage:

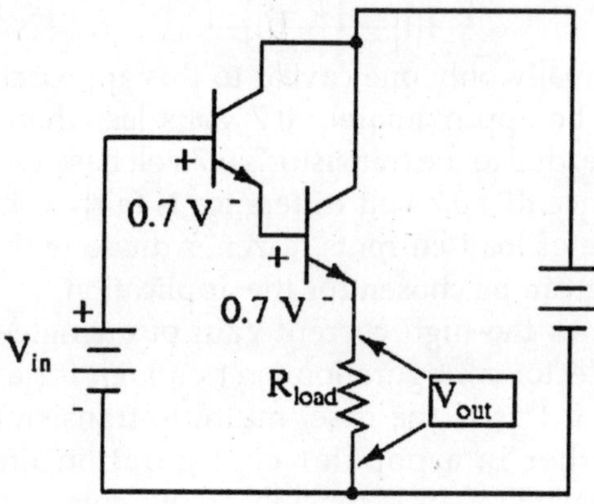

$$V_{out} = V_{in} - 1.4$$

Darlington pairs may be purchased as discrete units (two transistors in the same package), or may be built up from a pair of individual transistors. Of course, if even more current gain is desired than what may be obtained with a pair, Darlington triplet or quadruplet assemblies may be constructed.

THE COMMON-BASE AMPLIFIER

The final transistor amplifier configuration we need to study is the *common-base*. This configuration is more complex than the other two, and is less common due to its strange operating characteristics.

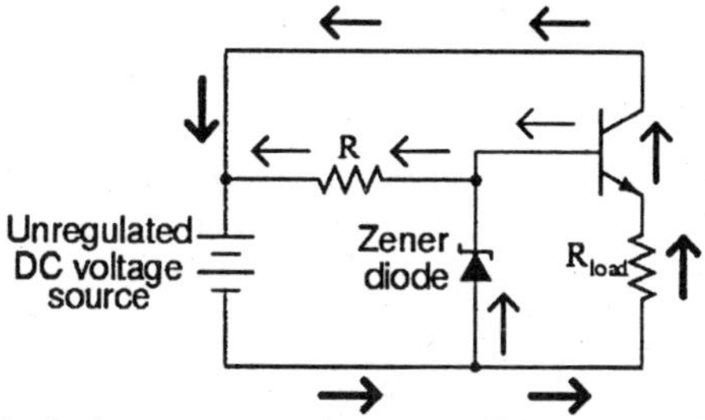

It is called the *common-base* configuration because (DC power source aside), the signal source and the load share the base of the transistor as a common connection point:

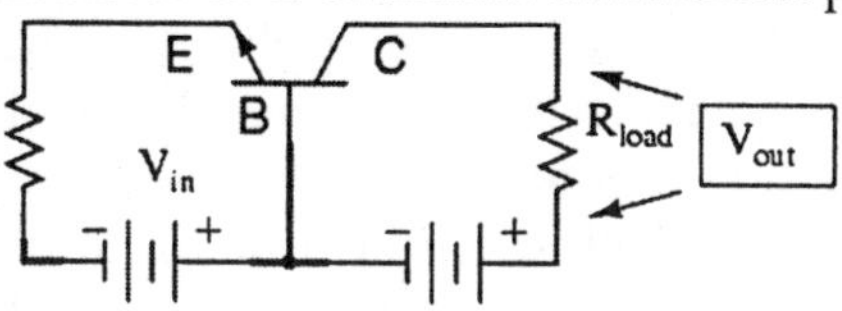

Perhaps the most striking characteristic of this configuration is that the input signal source must carry the full emitter current of the transistor, as indicated by the heavy arrows in the first illustration.

As we know, the emitter current is greater than any other current in the transistor, being the sum of base and collector currents.

In the last two amplifier configurations, the signal source was connected to the base lead of the transistor, thus handling the *least* current possible.

Because the input current exceeds all other currents in the circuit, including the output current, the current gain of this amplifier is actually *less than 1* (notice how R_{load} is connected to the collector, thus carrying slightly less current than the signal source). In other words, it *attenuates* current rather than *amplifying* it.

With common-emitter and common-collector amplifier configurations, the transistor parameter most closely associated with gain was β. In the common-base circuit, we follow another basic transistor parameter: the ratio between collector current and emitter current, which is a fraction always less than 1. This fractional value for any transistor is called the *alpha* ratio, or á ratio.

BIASING TECHNIQUES

In the common-emitter, we saw analysis where the output waveform resembled a half-wave rectified shape: only half of the input waveform was reproduced, with the other half being completely cut off. Since our purpose at that time was to reproduce the entire wave shape, this constituted a problem. The solution to this problem was to add a small bias voltage to the amplifier input so that the transistor stayed in active mode throughout the entire wave cycle. This addition was called a *bias voltage*.

There are applications, though, where a half-wave output is not problematic. In fact, some applications may *necessitate* this very type of amplification. Because it is possible to operate an amplifier in modes other than full-wave reproduction, and because there are specific applications requiring different ranges of reproduction, it is useful to describe the degree to which an amplifier reproduces the input waveform by designating it according to *class*. Amplifier class operation is categorized by means of alphabetical letters: A, B, C, and AB.

Class A operation is where the entire input waveform is faithfully reproduced. Although I didn't introduce this concept back in the common-emitter section, this is what we were hoping to attain in our simulations. Class A operation can only be obtained when the transistor spends its entire time in the active mode, never reaching either cutoff or saturation. To achieve this, sufficient DC bias voltage is usually set at the level necessary to drive the transistor exactly halfway between cutoff and saturation. This way, the AC input signal will be perfectly "centered" between the amplifier's high and low signal limit levels.

Class B operation is what we had the first time an AC signal was applied to the common-emitter amplifier with no DC bias voltage. The transistor spent half its time in active mode and the other half in cutoff with the input voltage too low (or even of the wrong polarity!) to forward-bias its base-emitter junction.

By itself, an amplifier operating in class B mode is not very useful. In most circumstances, the severe distortion introduced into the waveshape by eliminating half of it would be unacceptable. However, class B operation is a useful mode of biasing if two amplifiers are operated as a *push-pull* pair, each amplifier handling only half of the waveform at a time:

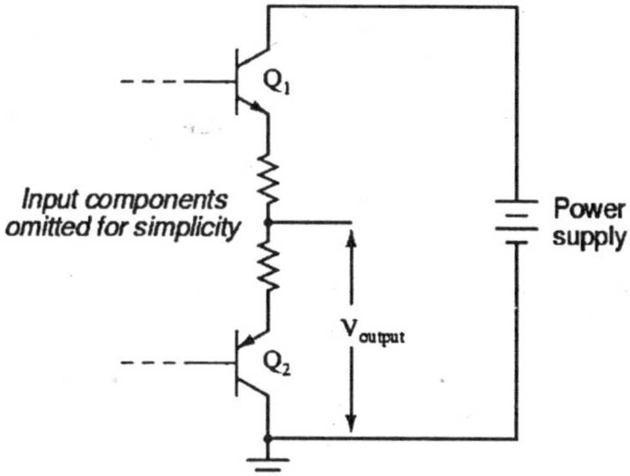

Transistor Q_1 "pushes" (drives the output voltage in a positive direction with respect to ground), while transistor Q_2 "pulls" the output voltage (in a negative direction, toward 0 volts with respect to ground). Individually, each of these transistors is operating in class B mode, active only for one-half of the input waveform cycle. Together, however, they function as a team to produce an output waveform identical in shape to the input waveform.

A decided advantage of the class B (push-pull) amplifier design over the class A design is greater output power capability. With a class A design, the transistor dissipates a lot of energy in the form of heat because it never stops conducting current. At all points in the wave cycle it is in the active (conducting) mode, conducting substantial current and dropping substantial voltage. This means there is substantial power dissipated by the transistor throughout the cycle. In a class B design, each transistor spends half the time in cutoff mode, where it dissipates zero power (zero current = zero power dissipation). This gives each transistor a time to "rest" and cool while the other transistor carries the burden of the load. Class A amplifiers are simpler in design, but tend to be limited to low-power signal applications for the simple reason of transistor heat dissipation.

There is another class of amplifier operation known as *class AB*, which is somewhere between class A and class B: the transistor spends more than 50% but less than 100% of the time conducting current.

If the input signal bias for an amplifier is slightly negative (opposite of the bias polarity for class A operation), the output waveform will be further "clipped" than it was with class B biasing, resulting in an operation where the transistor spends the majority of the time in cutoff mode:

At first, this scheme may seem utterly pointless. After all, how useful could an amplifier be if it clips the waveform as badly as this? If the output is used directly with no conditioning of any kind, it would indeed be of questionable utility. However, with the application of a tank circuit (parallel resonant inductor-capacitor combination) to the output, the occasional output surge produced by the amplifier can set in motion a higher-frequency oscillation maintained by the tank circuit. This may be likened to a machine where a heavy flywheel is given an occasional "kick" to keep it spinning:

Called *class* C operation, this scheme also enjoys high power efficiency due to the fact that the transistor(s) spend the vast majority of time in the cutoff mode, where they dissipate zero power. The rate of output waveform decay (decreasing oscillation amplitude between "kicks" from the amplifier) is exaggerated here for the benefit of illustration. Because of the tuned tank circuit on the output, this type of circuit is usable only for amplifying signals of definite, fixed frequency.

Another type of amplifier operation, significantly different from Class A, B, AB, or C, is called *Class D*. It is not obtained by applying a specific measure of bias voltage as are the other classes of operation, but requires a radical re-design of the amplifier circuit itself. A class D amplifier reproduces the profile of the input voltage waveform by generating a rapidly-pulsing squarewave output. The duty cycle of this output waveform (time "on" versus total cycle time) varies with the instantaneous amplitude of the input signal. The following plots demonstrate this principle:

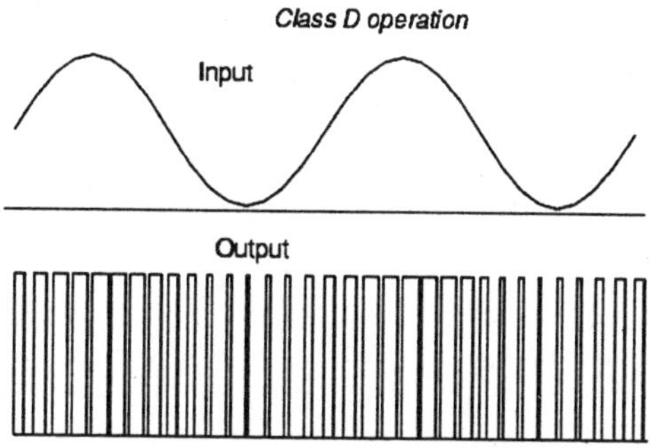

The greater the instantaneous voltage of the input signal, the greater the duty cycle of the output squarewave pulse. If there can be any goal stated of the class D design, it is to avoid active-mode transistor operation. Since the output transistor of a class D amplifier is never in the active mode, only cutoff or saturated, there will be little heat energy dissipated by it.

This results in very high power efficiency for the amplifier. Of course, the disadvantage of this strategy is the overwhelming presence of harmonics on the output. Fortunately, since these harmonic frequencies are typically much greater than the frequency of the input signal, they can be filtered out by a low-pass filter with relative ease, resulting in an output more closely resembling the original input signal waveform.

Class D technology is typically seen where extremely high power levels and relatively low frequencies are encountered, such as in industrial inverters (devices converting DC into AC power to run motors and other large devices) and high-performance audio amplifiers.

A term you will likely come across in your studies of electronics is something called *quiescent,* which is a modifier designating the normal, or zero input signal, condition of a circuit. Quiescent current, for example, is the amount of current in a circuit with zero input signal voltage applied. Bias voltage in a transistor circuit forces the transistor to operate at a different level of collector current with zero input signal voltage than it would without that bias voltage. Therefore, the amount of bias in an amplifier circuit determines its quiescent values.

In a class A amplifier, the quiescent current should be exactly half of its saturation value (halfway between saturation and cutoff, cutoff by definition being zero). Class B and class C amplifiers have quiescent current values of zero, since they are supposed to be cutoff with no signal applied. Class AB amplifiers have very low quiescent current values, just above cutoff.

To illustrate this graphically, a "load line" is sometimes plotted over a transistor's characteristic curves to illustrate its range of operation while connected to a load resistance of specific value:

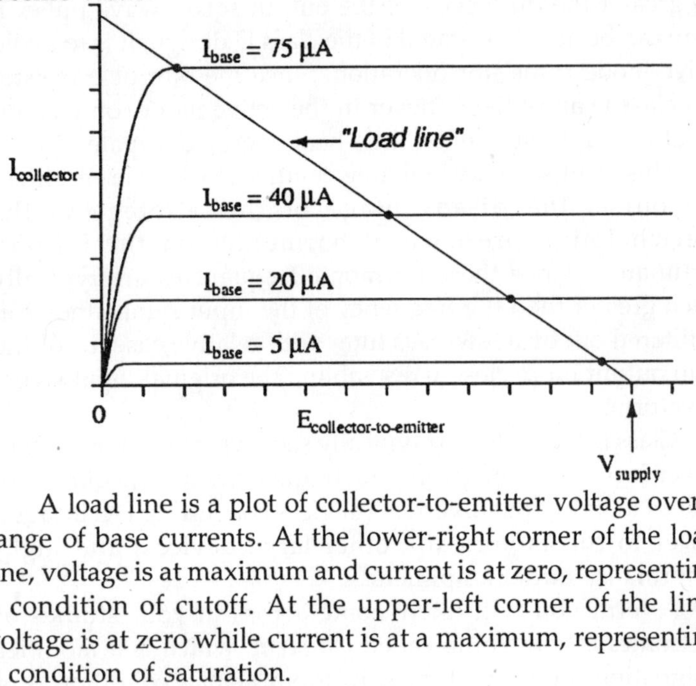

A load line is a plot of collector-to-emitter voltage over a range of base currents. At the lower-right corner of the load line, voltage is at maximum and current is at zero, representing a condition of cutoff. At the upper-left corner of the line, voltage is at zero while current is at a maximum, representing a condition of saturation.

Dots marking where the load line intersects the various transistor curves represent realistic operating conditions for those base currents given.

Quiescent operating conditions may be shown on this type of graph in the form of a single dot along the load line. For a class A amplifier, the quiescent point will be in the middle of the load line, like this:

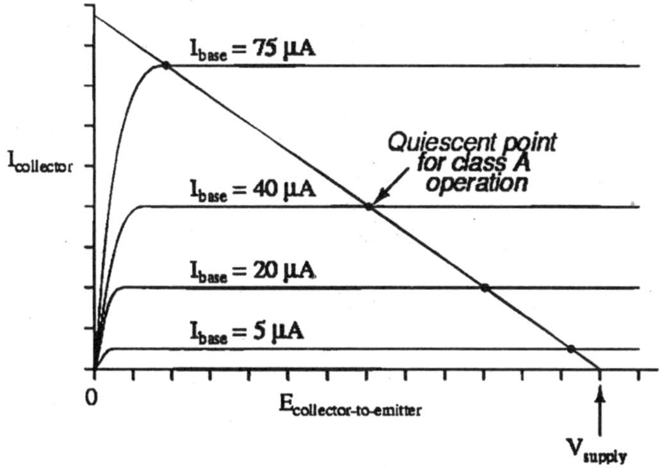

In this illustration, the quiescent point happens to fall on the curve representing a base current of 40 µA. If we were to change the load resistance in this circuit to a greater value, it would affect the slope of the load line, since a greater load resistance would limit the maximum collector current at saturation, but would not change the collector-emitter voltage at cutoff. Graphically, the result is a load line with a different upper-left point and the same lower-right point:

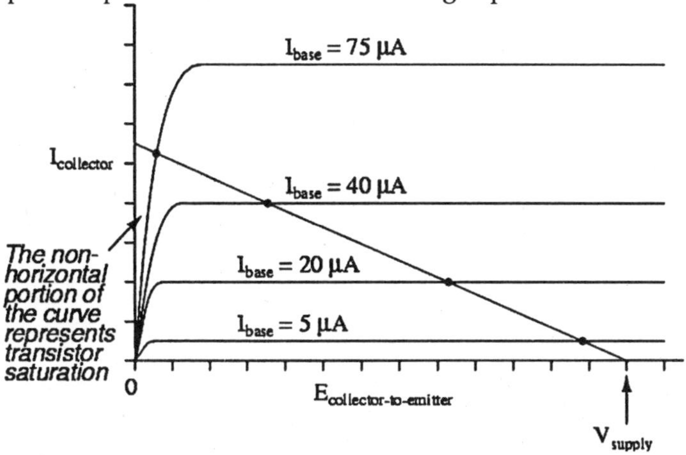

Note how the new load line doesn't intercept the 75 µA curve along its flat portion as before. This is very important to realize because the non-horizontal portion of a characteristic curve represents a condition of saturation. Having the load line intercept the 75 μA curve outside of the curve's horizontal range means that the amplifier will be saturated at that amount of base current. Increasing the load resistor value is what caused the load line to intercept the 75 μA curve at this new point, and it indicates that saturation will occur at a lesser value of base current than before.

With the old, lower-value load resistor in the circuit, a base current of 75 μA would yield a proportional collector current (base current multiplied by β). In the first load line graph, a base current of 75 μA gave a collector current almost twice what was obtained at 40 μA, as the β ratio would predict. Now, however, there is only a marginal increase in collector current between base current values of 75 μA and 40 μA, because the transistor begins to lose sufficient collector-emitter voltage to continue to regulate collector current.

In order to maintain linear (no-distortion) operation, transistor amplifiers shouldn't be operated at points where the transistor will saturate; that is, in any case where the load line will not potentially fall on the horizontal portion of a collector current curve. In this case, we'd have to add a few more curves to the graph before we could tell just how far we could "push" this transistor with increased base currents before it saturates.

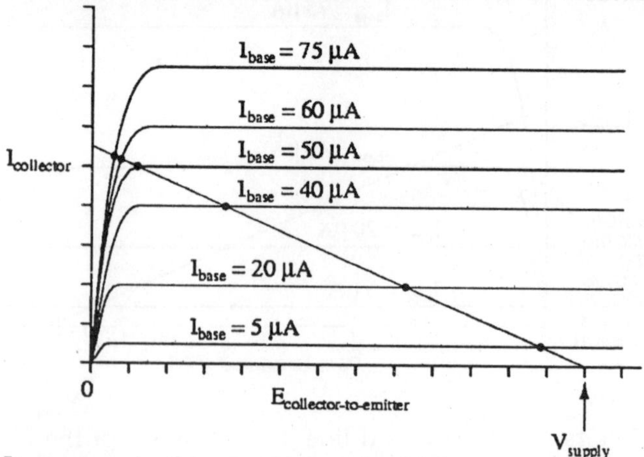

It appears in this graph that the highest-current point on the load line falling on the straight portion of a curve is the point on the 50 μA curve. This new point should be considered the maximum allowable input signal level for class A operation.

Also for class A operation, the bias should be set so that the quiescent point is halfway between this new maximum point and cutoff:

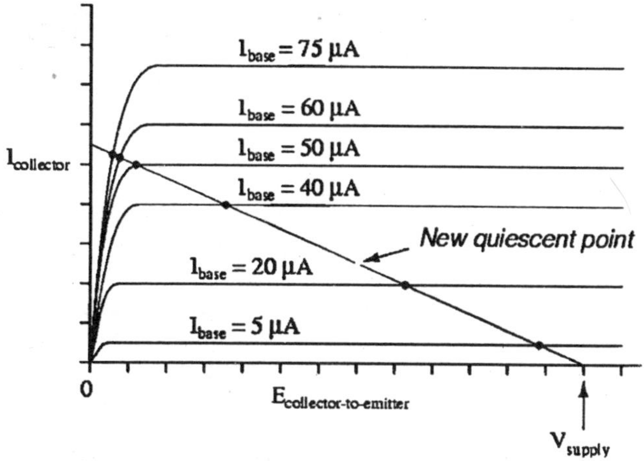

Now that we know a little more about the consequences of different DC bias voltage levels, it is time to investigate practical biasing techniques. So far, I've shown a small DC voltage source (battery) connected in series with the AC input signal to bias the amplifier for whatever desired class of operation.

In real life, the connection of a precisely-calibrated battery to the input of an amplifier is simply not practical. Even if it were possible to customize a battery to produce just the right amount of voltage for any given bias requirement, that battery would not remain at its manufactured voltage indefinitely. Once it started to discharge and its output voltage drooped, the amplifier would begin to drift in the direction of class B operation.

Take this circuit, illustrated in the common-emitter section, for instance:

That 2.3 volt "V_{bias}" battery would not be practical to include in a real amplifier circuit. A far more practical method of obtaining bias voltage for this amplifier would be to develop the necessary 2.3 volts using a voltage divider network connected across the 15 volt battery.

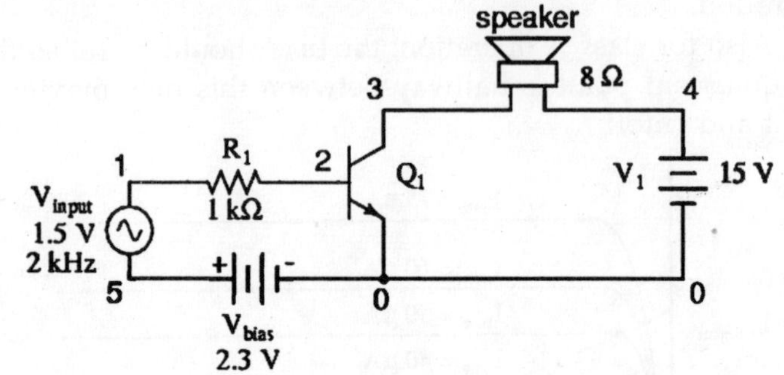

After all, the 15 volt battery is already there by necessity, and voltage divider circuits are very easy to design and build. Let's see how this might look:

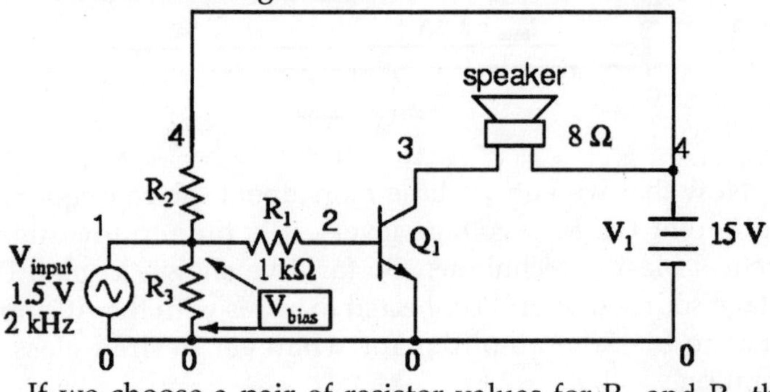

If we choose a pair of resistor values for R_2 and R_3 that will produce 2.3 volts across R_3 from a total of 15 volts (such as 8466 Ω for R_2 and 1533 Ω for R_3), we should have our desired value of 2.3 volts between base and emitter for biasing with no signal input. The only problem is, this circuit configuration places the AC input signal source directly in parallel with R_3 of our voltage divider. This is not acceptable, as the AC source will tend to overpower any DC voltage dropped across R_3. Parallel components *must* have the same voltage, so if an AC voltage source is directly connected across one resistor of a DC voltage divider, the AC source will "win" and there will be no DC bias voltage added to the signal.

One way to make this scheme work, although it may not be obvious *why* it will work, is to place a *coupling capacitor* between the AC voltage source and the voltage divider like this:

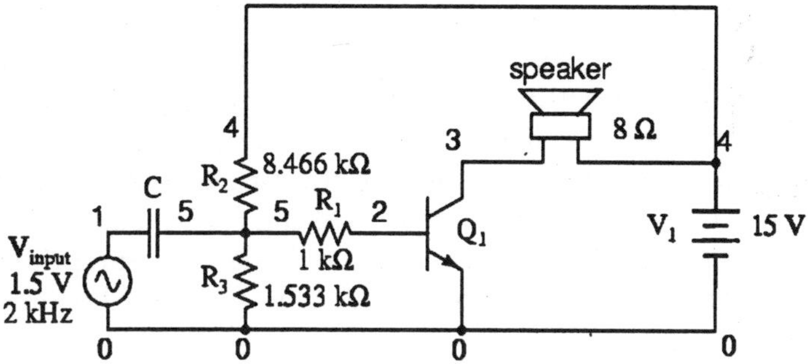

The capacitor forms a high-pass filter between the AC source and the DC voltage divider, passing almost all of the AC signal voltage on to the transistor while blocking all DC voltage from being shorted through the AC signal source. This makes much more sense if you understand the superposition theorem and how it works. According to superposition, any linear, bilateral circuit can be analyzed in a piecemeal fashion by only considering one power source at a time, then algebraically adding the effects of all power sources to find the final result. If we were to separate the capacitor and R_2—R_3 voltage divider circuit from the rest of the amplifier, it might be easier to understand how this superposition of AC and DC would work.

With only the AC signal source in effect, and a capacitor with an arbitrarily low impedance at signal frequency, almost all the AC voltage appears across R_3:

With only the DC source in effect, the capacitor appears to be an open circuit, and thus neither it nor the shorted AC signal source will have any effect on the operation of the R_2—R_3 voltage divider:

Combining these two separate analyses, we get a superposition of (almost) 1.5 volts AC and 2.3 volts DC, ready to be connected to the base of the transistor:

Notice that there is substantial distortion in the output waveform here: the sine wave is being clipped during most of the input signal's negative half-cycle. This tells us the transistor is entering into cutoff mode when it shouldn't (I'm assuming a goal of class A operation as before). Why is this? This new biasing technique should give us exactly the same amount of DC bias voltage as before, right?

With the capacitor and R_2—R_3 resistor network unloaded, it will provide exactly 2.3 volts worth of DC bias. However, once we connect this network to the transistor, it is no longer unloaded. Current drawn through the base of the transistor will load the voltage divider, thus reducing the DC bias voltage available for the transistor. Using the diode-regulating diode transistor model to illustrate, the bias problem becomes evident:

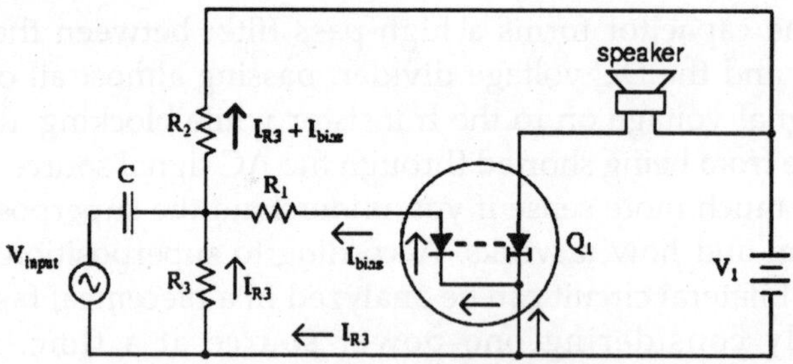

A voltage divider's output depends not only on the size of its constituent resistors, but also on how much current is being divided away from it through a load. In this case, the base-emitter PN junction of the transistor is a load that decreases the DC voltage dropped across R_3, due to the fact that the bias current joins with R_3's current to go through R_2, upsetting the divider ratio formerly set by the resistance values of R_2 and R_3. In order to obtain a DC bias voltage of 2.3 volts, the values of R_2 and/or R_3 must be adjusted to compensate for the effect of base current loading. In this case, we want to *increase* the DC voltage dropped across R_3, so we can lower the value of R_2, raise the value of R_3, or both.

INPUT AND OUTPUT COUPLING

To overcome the challenge of creating necessary DC bias voltage for an amplifier's input signal without resorting to the insertion of a battery in series with the AC signal source, we used a voltage divider connected across the DC power source. To make this work in conjunction with an AC input signal, we "coupled" the signal source to the divider through a capacitor, which acted as a high-pass filter. With that filtering in place, the low impedance of the AC signal source couldn't "short out" the DC voltage dropped across the bottom resistor of the voltage divider. A simple solution, but not without any disadvantages.

Most obvious is the fact that using a high-pass filter capacitor to couple the signal source to the amplifier means that the amplifier can only amplify AC signals. A steady, DC voltage applied to the input would be blocked by the coupling capacitor just as much as the voltage divider bias voltage is blocked from the input source. Furthermore, since capacitive reactance is frequency-dependent, lower-frequency AC signals will not be amplified as much as higher-frequency signals. Non-sinusoidal signals will tend to be distorted, as the capacitor responds differently to each of the signal's constituent harmonics. An extreme example of this would be a low-frequency square-wave signal:

Incidentally, this same problem occurs when oscilloscope inputs are set to the "AC coupling" mode. In this mode, a coupling capacitor is inserted in series with the measured voltage signal to eliminate any vertical offset of the displayed waveform due to DC voltage combined with the signal. This works fine when the AC component of the measured signal is of a fairly high frequency, and the capacitor offers little impedance to the signal.

However, if the signal is of a low frequency, and/or contains considerable levels of harmonics over a wide frequency range, the oscilloscope's display of the waveform will not be accurate.

In applications where the limitations of capacitive coupling would be intolerable, another solution may be used: *direct coupling*. Direct coupling avoids the use of capacitors or any other frequency-dependent coupling component in favor of resistors. A direct-coupled amplifier circuit might look something like this:

With no capacitor to filter the input signal, this form of coupling exhibits no frequency dependence.

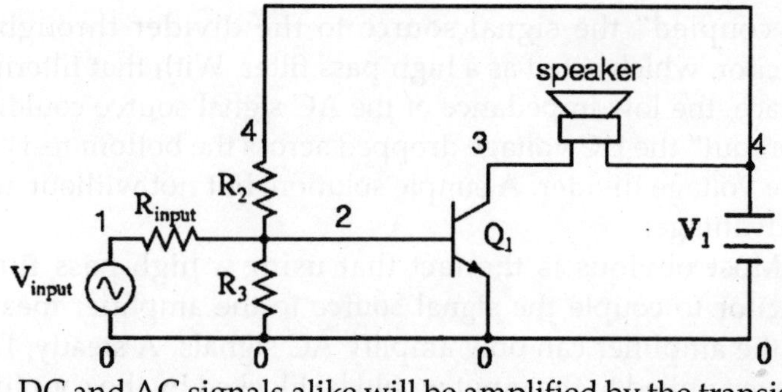

DC and AC signals alike will be amplified by the transistor with the same gain (the transistor itself may tend to amplify some frequencies better than others, but that is another subject entirely!).

If direct coupling works for DC as well as for AC signals, then why use capacitive coupling for *any* application? One reason might be to avoid any *unwanted* DC bias voltage naturally present in the signal to be amplified. Some AC signals may be superimposed on an uncontrolled DC voltage right from the source, and an uncontrolled DC voltage would make reliable transistor biasing impossible. The high-pass filtering offered by a coupling capacitor would work well here to avoid biasing problems.

Another reason to use capacitive coupling rather than direct is its relative lack of signal attenuation. Direct coupling through a resistor has the disadvantage of diminishing, or attenuating, the input signal so that only a fraction of it reaches the base of the transistor. In many applications, some attenuation is necessary anyway to prevent normal signal levels from "overdriving" the transistor into cutoff and saturation, so any attenuation inherent to the coupling network is useful anyway. However, some applications require that there be *no* signal loss from the input connection to the transistor's base for maximum voltage gain, and a direct coupling scheme with a voltage divider for bias simply won't suffice.

So far, we've discussed a couple of methods for coupling an *input* signal to an amplifier, but haven't addressed the issue of coupling an amplifier's *output* to a load. The example circuit used to illustrate input coupling will serve well to illustrate the issues involved with output coupling.

In our example circuit, the load is a speaker. Most speakers are electromagnetic in design: that is, they use the force generated by an lightweight electromagnet coil suspended within a strong permanent-magnet field to move a thin paper or plastic cone, producing vibrations in the air which our ears interpret as sound. An applied voltage of one polarity moves the cone outward, while a voltage of the opposite polarity will move the cone inward.

To exploit cone's full freedom of motion, the speaker must receive true (unbiased) AC voltage. DC bias applied to the speaker coil tends to offset the cone from its natural center position, and this tends to limit the amount of back-and-forth motion it can sustain from the applied AC voltage without overtraveling. However, our example circuit applies a varying voltage of only *one* polarity across the speaker, because the speaker is connected in series with the transistor which can only conduct current one way. This situation would be unacceptable in the case of any high-power audio amplifier.

Somehow we need to isolate the speaker from the DC bias of the collector current so that it only receives AC voltage. One way to achieve this goal is to couple the transistor collector circuit to the speaker through a transformer:

Voltage induced in the secondary (speaker-side) of the transformer will be strictly due to *variations* in collector current, because the mutual inductance of a transformer only works on *changes* in winding current. In other words, only the AC portion of the collector current signal will be coupled to the secondary side for powering the speaker. The speaker will "see" true alternating current at its terminals, without any DC bias.

Transformer output coupling works, and has the added benefit of being able to provide impedance matching between the transistor circuit and the speaker coil with custom winding ratios. However, transformers tend to be large and heavy, especially for high-power applications. Also, it is difficult to engineer a transformer to handle signals over a wide range of frequencies, which is almost always required for audio applications.

To make matters worse, DC current through the primary winding adds to the magnetization of the core in one polarity only, which tends to make the transformer core saturate more easily in one AC polarity cycle than the other. This problem is reminiscent of having the speaker directly connected in series with the transistor: a DC bias current tends to limit how much output signal amplitude the system can handle without distortion. Generally, though, a transformer can be designed to handle a lot more DC bias current than a speaker without running into trouble, so transformer coupling is still a viable solution in most cases.

Another method to isolate the speaker from DC bias in the output signal is to alter the circuit a bit and use a coupling capacitor in a manner similar to coupling the input signal to the amplifier:

This circuit resembles the more conventional form of common-emitter amplifier, with the transistor collector connected to the battery through a resistor.

The capacitor acts as a high-pass filter, passing most of the AC voltage to the speaker while blocking all DC voltage. Again, the value of this coupling capacitor is chosen so that its impedance at the expected signal frequency will be arbitrarily low.

The blocking of DC voltage from an amplifier's output, be it via a transformer or a capacitor, is useful not only in coupling an amplifier to a load, but also in coupling one amplfier to another amplifier.

Three-stage common-emitter amplifier

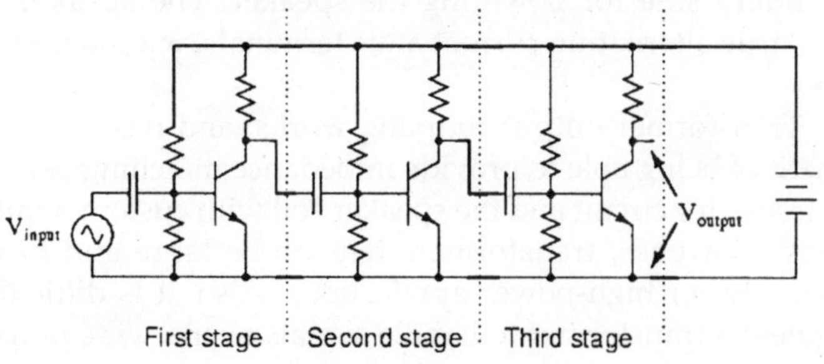

"Staged" amplifiers are often used to achieve higher power gains than what would be possible using a single transistor:

While it is possible to directly couple each stage to the next (via a resistor rather than a capacitor), this makes the whole amplifier *very* sensitive to variations in the DC bias voltage of the first stage, since that DC voltage will be amplified along with the AC signal until the last stage. In other words, the biasing of the first stage will affect the biasing of the second stage, and so on. However, if the stages are capacitively coupled as shown in the above illustration, the biasing of one stage has no effect on the biasing of the next, because DC voltage is blocked from passing on to the next stage.

Transformer coupling between amplifier stages is also a possibility, but less often seen due to some of the problems inherent to transformers mentioned previously. One notable exception to this rule is in the case of radio-frequency amplifiers where coupling transformers are typically small, have air cores (making them immune to saturation effects), and can be made part of a resonant circuit so as to block unwanted harmonic frequencies from passing on to subsequent stages. The use of resonant circuits assumes that the signal frequency remains constant, of course, but this is typically the case in radio circuitry. Also, the "flywheel" effect of LC tank circuits allows for class C operation for high efficiency:

Three-stage tuned (RF) amplifier

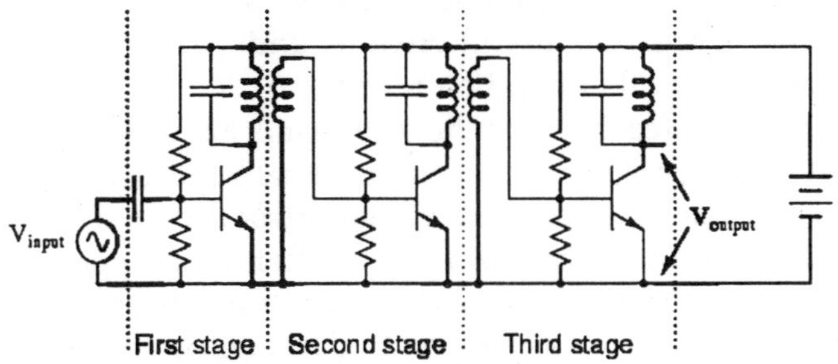

Having said all this, it must be mentioned that it *is* possible to use direct coupling within a multi-stage transistor amplifier circuit. In cases where the amplifier is expected to handle DC signals, this is the only alternative.

FEEDBACK

If some percentage of an amplifier's output signal is connected to the input, so that the amplifier amplifies part of its own output signal, we have what is known as *feedback*. Feedback comes in two varieties: *positive* (also called *regenerative*), and *negative* (also called *degenerative*). Positive feedback reinforces the direction of an amplifier's output voltage change, while negative feedback does just the opposite.

A familiar example of feedback happens in public-address ("PA") systems where someone holds the microphone too close to a speaker: a high-pitched "whine" or "howl" ensues, because the audio amplifier system is detecting and amplifying its own noise.

Specifically, this is an example of *positive* or *regenerative* feedback, as any sound detected by the microphone is amplified and turned into a louder sound by the speaker, which is then detected by the microphone again, and so on.

One might wonder what possible benefit feedback is to an amplifier circuit, given such an annoying example as PA system "howl." If we introduce positive, or regenerative, feedback into an amplifier circuit, it has the tendency of creating and sustaining oscillations, the frequency of which determined by the values of components handling the feedback signal from output to input. This is one way to make an *oscillator* circuit to produce AC from a DC power supply. Oscillators are very useful circuits, and so feedback has a definite, practical application for us.

Negative feedback, on the other hand, has a "dampening" effect on an amplifier: if the output signal happens to increase in magnitude, the feedback signal introduces a decreasing influence into the input of the amplifier, thus opposing the change in output signal. While positive feedback drives an amplifier circuit toward a point of instability (oscillations), negative feedback drives it the opposite direction: toward a point of stability.

An amplifier circuit equipped with some amount of negative feedback is not only more stable, but it tends to distort the input waveform to a lesser degree and is generally capable of amplifying a wider range of frequencies. The tradeoff for these advantages (there just *has* to be a disadvantage to negative feedback, right?) is decreased gain. If a portion of an amplifier's output signal is "fed back" to the input in such a way as to oppose any changes in the output, it will require a greater input signal amplitude to drive the amplifier's output to the same amplitude as before. This constitutes a decreased gain. However, the advantages of stability, lower distortion, and greater bandwidth are worth the tradeoff in reduced gain for many applications.

Let's examine a simple amplifier circuit and see how we might introduce negative feedback into it:

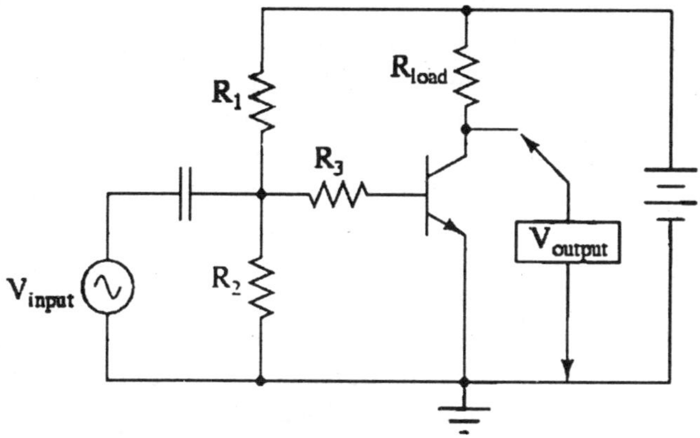

The amplifier configuration shown here is a common-emitter, with a resistor bias network formed by R_1 and R_2. The capacitor couples V_{input} to the amplifier so that the signal source doesn't have a DC voltage imposed on it by the R_1/R_2 divider network. Resistor R_3 serves the purpose of controlling voltage gain. We could omit if for maximum voltage gain, but since base resistors like this are common in common-emitter amplifier circuits, we'll keep it in this schematic.

Like all common-emitter amplifiers, this one *inverts* the input signal as it is amplified. In other words, a positive-going input voltage causes the output voltage to decrease, or go in the direction of negative, and vice versa. If we were to examine the waveforms with oscilloscopes, it would look something like this:

Because the output is an inverted, or mirror-image, reproduction of the input signal, any connection between the output (collector) wire and the input (base) wire of the transistor will result in *negative* feedback:

The resistances of R_1, R_2, R_3, and $R_{feedback}$ function together as a signal-mixing network so that the voltage seen at the base of the transistor (in reference to ground) is a weighted average of the input voltage and the feedback voltage, resulting in signal of reduced amplitude going into the transistor.

As a result, the amplifier circuit will have reduced voltage gain, but improved linearity (reduced distortion) and increased bandwidth.

A resistor connecting collector to base is not the only way to introduce negative feedback into this amplifier circuit, though. Another method, although more difficult to understand at first, involves the placement of a resistor between the transistor's emitter terminal and circuit ground, like this:

This new feedback resistor drops voltage proportional to the emitter current through the transistor, and it does so in such a way as to oppose the input signal's influence on the base-emitter junction of the transistor. Let's take a closer look at the emitter-base junction and see what difference this new resistor makes:

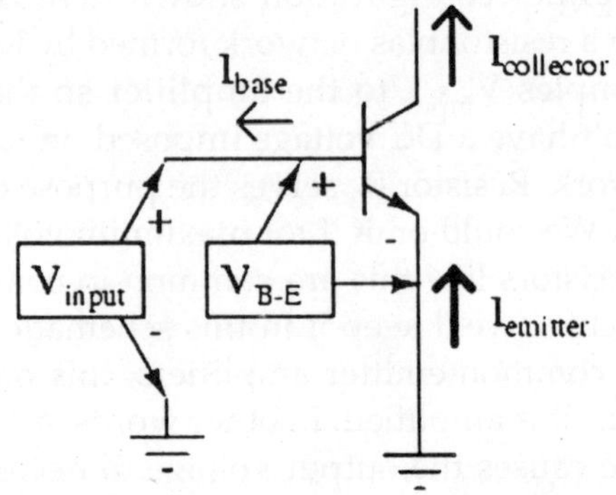

With no feedback resistor connecting the emitter to ground, whatever level of input signal (V_{input}) makes it through the coupling capacitor and $R_1/R_2/R_3$ resistor network will be impressed directly across the base-emitter junction as the transistor's input voltage (V_{B-E}). In other words, with no feedback resistor, V_{B-E} equals V_{input}. Therefore, if V_{input} increases by 100 mV, then V_{B-E} likewise increases by 100 mV: a change in one is the same as a change in the other, since the two voltages are equal to each other.

Now let's consider the effects of inserting a resistor ($R_{feedback}$) between the transistor's emitter lead and ground:

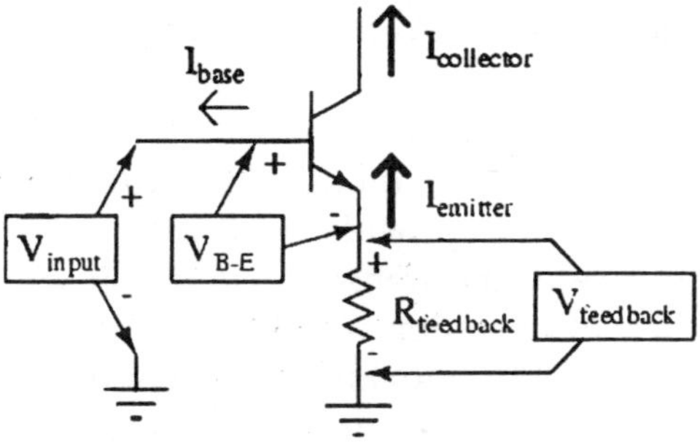

Note how the voltage dropped across $R_{feedback}$ adds with V_{B-E} to equal V_{input}. With $R_{feedback}$ in the $V_{input} - V_{B-E}$ loop, V_{B-E} will no longer be equal to V_{input}. We know that $R_{feedback}$ will drop a voltage proportional to emitter current, which is in turn controlled by the base current, which is in turn controlled by the voltage dropped across the base-emitter junction of the transistor (V_{B-E}). Thus, if V_{input} were to increase in a positive direction, it would increase V_{B-E}, causing more base current, causing more collector (load) current, causing more emitter current, and causing more feedback voltage to be dropped across $R_{feedback}$.

This increase of voltage drop across the feedback resistor, though, *subtracts* from V_{input} to reduce the V_{B-E}, so that the actual voltage increase for V_{B-E} will be less than the voltage increase of V_{input}. No longer will a 100 mV increase in V_{input} result in a full 100 mV increase for V_{B-E}, because the two voltages are *not* equal to each other. –

Consequently, the input voltage has less control over the

transistor than before, and the voltage gain for the amplifier is reduced: just what we expected from negative feedback.

In practical common-emitter circuits, negative feedback isn't just a luxury; its a necessity for stable operation. In a perfect world, we could build and operate a common-emitter transistor amplifier with no negative feedback, and have the full amplitude of V_{input} impressed across the transistor's base-emitter junction.

This would give us a large voltage gain. Unfortunately, though, the relationship between base-emitter voltage and base-emitter current changes with temperature, as predicted by the "diode equation." As the transistor heats up, there will be less of a forward voltage drop across the base-emitter junction for any given current.

This causes a problem for us, as the R_1/R_2 voltage divider network is designed to provide the correct quiescent current through the base of the transistor so that it will operate in whatever class of operation we desire (in this example, I've shown the amplifier working in class-A mode). If the transistor's voltage/current relationship changes with temperature, the amount of DC bias voltage necessary for the desired class of operation will change. In this case, a hot transistor will draw more bias current for the same amount of bias voltage, making it heat up even more, drawing even more bias current. The result, if unchecked, is called *thermal runaway*.

Common-collector amplifiers, however, do not suffer from thermal runaway. Why is this? The answer has everything to do with negative feedback:

Note that the common-collector amplifier has its load resistor placed in exactly the same spot as we had the $R_{feedback}$ resistor in the last circuit: between emitter and ground. This means that the only voltage impressed across the transistor's base-emitter junction is the *difference* between V_{input} and V_{output}, resulting in a very low voltage gain (usually close to 1 for a common-collector amplifier). Thermal runaway is impossible for this amplifier: if base current happens to increase due to transistor heating, emitter current will likewise increase, dropping more voltage across the load, which in turn *subtracts* from V_{input} to reduce the amount of voltage dropped between base and emitter. In other words, the negative feedback afforded by placement of the load resistor makes the problem of thermal runaway *self-correcting*. In exchange for a greatly reduced voltage gain, we get superb stability and immunity from thermal runaway.

By adding a "feedback" resistor between emitter and ground in a common-emitter amplifier, we make the amplifier behave a little less like an "ideal" common-emitter and a little more like a common-collector. The feedback resistor value is typically quite a bit less than the load, minimizing the amount of negative feedback and keeping the voltage gain fairly high.

Another benefit of negative feedback, seen clearly in the common-collector circuit, is that it tends to make the voltage gain of the amplifier less dependent on the characteristics of the transistor. Note that in a common-collector amplifier, voltage gain is nearly equal to unity (1), regardless of the transistor's β.

This means, among other things, that we could replace the transistor in a common-collector amplifier with one having a different β and not see any significant changes in voltage gain. In a common-emitter circuit, the voltage gain is highly dependent on β. If we were to replace the transistor in a common-emitter circuit with another of differing β, the voltage gain for the amplifier would change significantly. In a common-emitter amplifier equipped with negative feedback, the voltage gain will still be dependent upon transistor β to some degree, but not as much as before, making the circuit more predictable despite variations in transistor β.

The fact that we have to introduce negative feedback into a common-emitter amplifier to avoid thermal runaway is an unsatisfying solution. It would be nice, after all, to avoid thermal runaway without having to suppress the amplifier's inherently high voltage gain.

A best-of-both-worlds solution to this dilemma is available to us if we closely examine the nature of the problem: the voltage gain that we have to minimize in order to avoid thermal runaway is the *DC* voltage gain, not the *AC* voltage gain. After all, it isn't the AC input signal that fuels thermal runaway: its the DC bias voltage required for a certain class of operation: that quiescent DC signal that we use to "trick" the transistor (fundamentally a DC device) into amplifying an AC signal.

We can suppress DC voltage gain in a common-emitter amplifier circuit without suppressing AC voltage gain if we figure out a way to make the negative feedback function with DC only. That is, if we only feed back an inverted DC signal from output to input, but not an inverted AC signal.

The $R_{feedback}$ emitter resistor provides negative feedback by dropping a voltage proportional to load current. In other words, negative feedback is accomplished by inserting an impedance into the emitter current path. If we want to feed back DC but not AC, we need an impedance that is high for DC but low for AC. What kind of circuit presents a high impedance to DC but a low impedance to AC? A high-pass filter, of course!

By connecting a capacitor in parallel with the feedback resistor, we create the very situation we need: a path from emitter to ground that is easier for AC than it is for DC:

The new capacitor "bypasses" AC from the transistor's emitter to ground, so that no appreciable AC voltage will be dropped from emitter to ground to "feed back" to the input and suppress voltage gain. Direct current, on the other hand, cannot go through the bypass capacitor, and so must travel through the feedback resistor, dropping a DC voltage between emitter and ground which lowers the DC voltage gain and stabilizes the amplifier's DC response, preventing thermal runaway. Because we want the reactance of this capacitor (X_C) to be as low as possible, C_{bypass} should be sized relatively large. Because the polarity across this capacitor will never change, it is safe to use a polarized (electrolytic) capacitor for the task.

Another approach to the problem of negative feedback reducing voltage gain is to use multi-stage amplifiers rather than single-transistor amplifiers. If the attenuated gain of a single transistor is insufficient for the task at hand, we can use more than one transistor to make up for the reduction caused by feedback. Here is an example circuit showing negative feedback in a three-stage common-emitter amplifier:

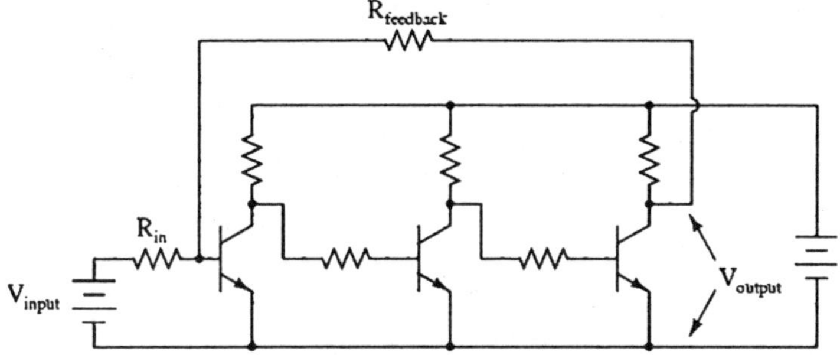

Note how there is but one "path" for feedback, from the final output to the input through a single resistor, $R_{feedback}$. Since each stage is a common-emitter amplifier — and thus inverting in nature — and there are an odd number of stages from input to output, the output signal will be inverted with respect to the input signal, and the feedback will be negative (degenerative). Relatively large amounts of feedback may be used without sacrificing voltage gain, because the three amplifier stages provide so much gain to begin with.

At first, this design philosophy may seem inelegant and perhaps even counter-productive. Isn't this a rather crude way to overcome the loss in gain incurred through the use of negative feedback, to simply recover gain by adding stage after stage? What is the point of creating a huge voltage gain using three transistor stages if we're just going to attenuate all that gain anyway with negative feedback?

The point, though perhaps not apparent at first, is increased predictability and stability from the circuit as a whole. If the three transistor stages are designed to provide an arbitrarily high voltage gain (in the tens of thousands, or greater) with no feedback, it will be found that the addition of negative feedback causes the overall voltage gain to become less dependent of the individual stage gains, and approximately equal to the simple ratio $R_{feedback}/R_{in}$.

The more voltage gain the circuit has (without feedback), the more closely the voltage gain will approximate $R_{feedback}/R_{in}$ once feedback is established. In other words, voltage gain in this circuit is fixed by the values of two resistors, and nothing more.

This advantage has profound impact on mass-production of electronic circuitry: if amplifiers of predictable gain may be constructed using transistors of widely varied β values, it makes the selection and replacement of components very easy and inexpensive. It also means the amplifier's gain varies little with changes in temperature. This principle of stable gain control through a high-gain amplifier "tamed" by negative feedback is elevated almost to an art form in electronic circuits called *operational amplifiers,* or *op-amps.*

CURRENT MIRRORS

An interesting and often-used circuit applying the bipolar junction transistor is the so-called *current mirror,* which serves as a simple current regulator, supplying nearly constant current to a load over a wide range of load resistances.

We know that in a transistor operating in its active mode, collector current is equal to base current multiplied by the ratio β. We also know that the ratio between collector current and emitter current is called á. Because collector current is equal to base current multiplied by β, and emitter current is the sum of the base and collector currents, á should be mathematically derivable from β. If you do the algebra, you'll find that á = β/(β+1) for any transistor.

We've seen already how maintaining a constant base current through an active transistor results in the regulation of collector current, according to the β ratio. Well, the á ratio works similarly: if emitter current is held constant, collector current will remain at a stable, regulated value so long as the transistor has enough collector-to-emitter voltage drop to maintain it in its active mode. Therefore, if we have a way of holding emitter current constant through a transistor, the transistor will work to regulate collector current at a constant value.

Remember that the base-emitter junction of a BJT is nothing more than a PN junction, just like a diode, and that the "diode equation" specifies how much current will go through a PN junction given forward voltage drop and junction temperature:

$$I_D = I_S \left(eq^{V_D/NkT} - 1\right)$$

Where,

I_D = Diode current in amps
I_S = Saturation current in amps (typically 1 × 10–12 amps).
e = Euler's constant (~ 2.71828.1828)
q = charge of electron (1.6×10^{-19} coulombs)
V_D = Voltage applied across diode in volts
N = "Nonideality" or "emisison" coefficient (typically between 1 and 2)
k = Boltzmann's constant (1.38 × 10–23)
T = Junction temperature in Kelvins.

If both junction voltage and temperature are held constant, then the PN junction current will likewise be constant. Following this rationale, if we were to hold the base-emitter voltage of a transistor constant, then its emitter current should likewise be constant, given a constant temperature:

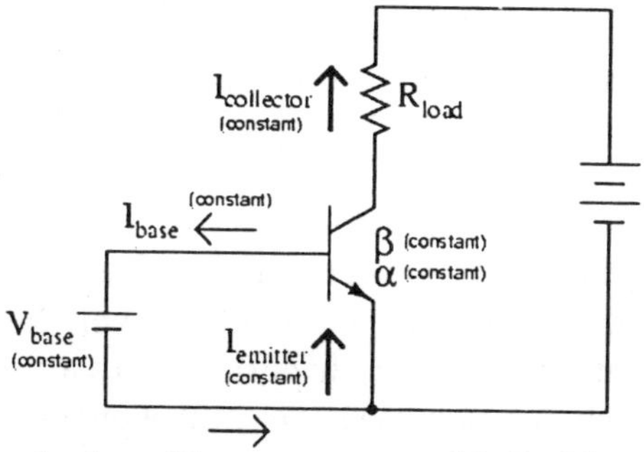

This constant emitter current, multiplied by a constant á ratio, gives a constant collector current through R_{load}, provided that there is enough battery voltage to keep the transistor in its active mode for any change in R_{load}'s resistance.

Maintaining a constant voltage across the transistor's base-emitter junction is easy: use a forward-biased diode to establish a constant voltage of approximately 0.7 volts, and connect it in parallel with the base-emitter junction:

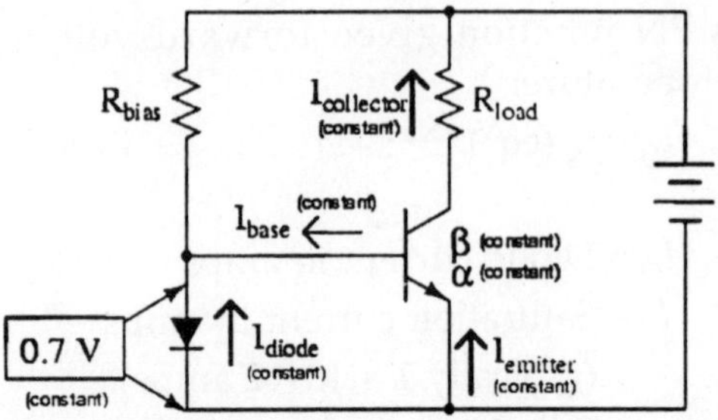

Now, here's where it gets interesting. The voltage dropped across the diode probably won't be 0.7 volts exactly. The exact amount of forward voltage dropped across it depends on the current through the diode, and the diode's temperature, all in accordance with the diode equation. If diode current is increased (say, by reducing the resistance of R_{bias}), its voltage drop will increase slightly, increasing the voltage drop across the transistor's base-emitter junction, which will increase the emitter current by the same proportion, assuming the diode's PN junction and the transistor's base-emitter junction are well-matched to each other. In other words, transistor emitter current will closely equal diode current at any given time. If you change the diode current by changing the resistance value of R_{bias}, then the transistor's emitter current will follow suit, because the emitter current is described by the same equation as the diode's, and both PN junctions experience the same voltage drop.

Remember, the transistor's collector current is almost equal to its emitter current, as the á ratio of a typical transistor is almost unity (1). If we have control over the transistor's emitter current by setting diode current with a simple resistor adjustment, then we likewise have control over the transistor's collector current. In other words, collector current mimics, or *mirrors*, diode current.

Current through resistor R_{load} is therefore a function of current set by the bias resistor, the two being nearly equal. This is the function of the current mirror circuit: to regulate current through the load resistor by conveniently adjusting the value of R_{bias}. It is very easy to create a set amount of diode current, as current through the diode is described by a simple equation: power supply voltage minus diode voltage (almost a constant value), divided by the resistance of R_{bias}.

To better match the characteristics of the two PN junctions (the diode junction and the transistor base-emitter junction), a transistor may be used in place of a regular diode, like this:

A current mirror circuit using two transistors

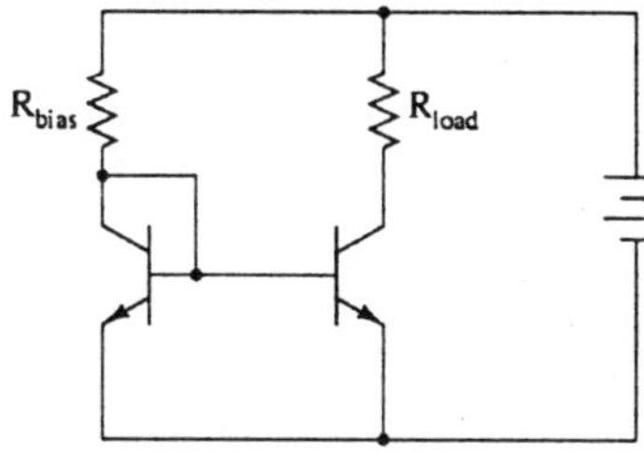

Because temperature is a factor in the "diode equation," and we want the two PN junctions to behave identically under all operating conditions, we should maintain the two transistors at exactly the same temperature. This is easily done using discrete components by gluing the two transistor cases back-to-back. If the transistors are manufactured together on a single chip of silicon (as a so-called *integrated circuit*, or *IC*), the designers should locate the two transistors very close to one another to facilitate heat transfer between them.

The current mirror circuit shown with two NPN transistors is sometimes called a *current-sinking* type, because the regulating transistor conducts current to the load *from ground* ("sinking" current), rather than *from the positive side of the battery* ("sourcing" current). If we wish to have a grounded load, and a *current sourcing* mirror circuit, we could use PNP transistors like this:

A ***current-sourcing*** *mirror circuit*

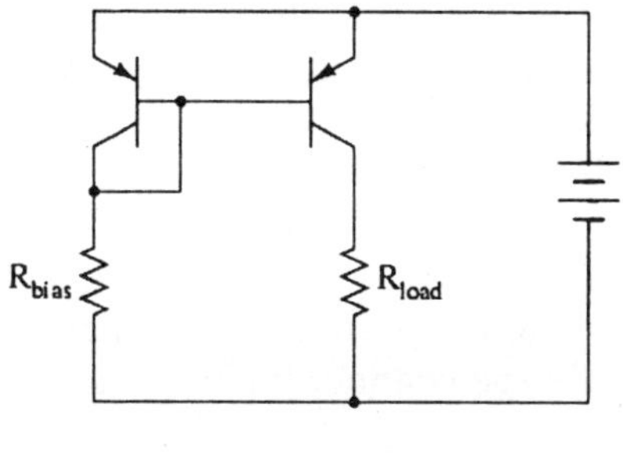

Chapter 4

Junction Field-effect Transistors

A *transistor* is a linear semiconductor device that controls current with the application of a lower-power electrical signal. Transistors may be roughly grouped into two major divisions: *bipolar* and *field-effect*. All field-effect transistors are *unipolar* rather than *bipolar* devices. That is, the main current through them is comprised either of electrons through an N-type semiconductor or holes through a P-type semiconductor. This becomes more evident when a physical diagram of the device is seen:

N-channel JFET

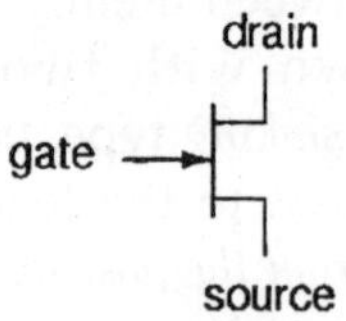

schematic symbol

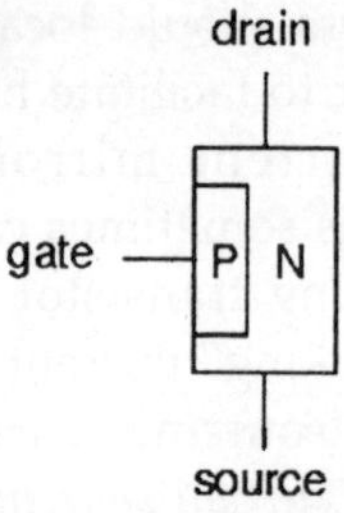

physical diagram

In a junction field-effect transistor, or JFET, the controlled current passes from source to drain, or from drain to source as the case may be. The controlling voltage is applied between the gate and source.

Note how the current does not have to cross through a PN junction on its way between source and drain: the path (called a *channel*) is an uninterrupted block of semiconductor material. In the image just shown, this channel is an N-type semiconductor. P-type channel JFETs are also manufactured:

***P-channel* JFET**

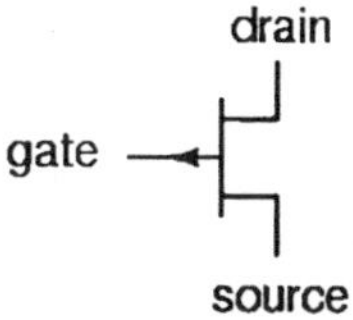

schematic symbol

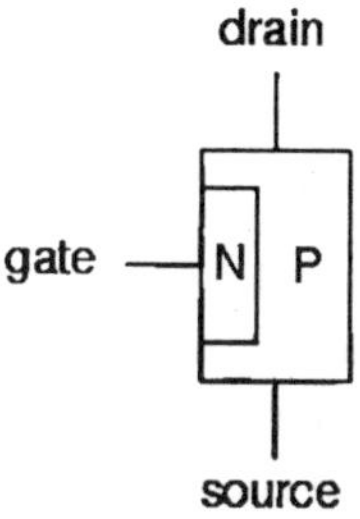

physical diagram

Generally, N-channel JFETs are more commonly used than P-channel. The reasons for this have to do with obscure details of semiconductor theory. The only practical difference between N- and P-channel JFETs you need to concern yourself with now is biasing of the PN junction formed between the gate material and the channel.

With no voltage applied between gate and source, the channel is a wide-open path for electrons to flow. However, if a voltage is applied between gate and source of such polarity that it reverse-biases the PN junction, the flow between source and drain connections becomes limited, or regulated, just as it was for bipolar transistors with a set amount of base current. Maximum gate-source voltage "pinches off" all current through source and drain, thus forcing the JFET into cutoff mode.

This behavior is due to the depletion region of the PN junction expanding under the influence of a reverse-bias voltage, eventually occupying the entire width of the channel if the voltage is great enough. This action may be likened to reducing the flow of a liquid through a flexible hose by squeezing it: with enough force, the hose will be constricted enough to completely block the flow.

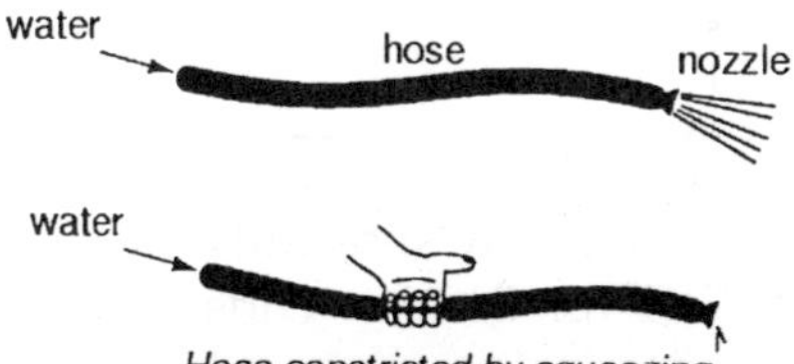

Hose constricted by squeezing, water flow reduced or stopped

Note how this operational behavior is exactly opposite of the bipolar junction transistor. Bipolar transistors are *normally-off* devices: no current through the base, no current through the collector or the emitter.

JFETs, on the other hand, are *normally-on* devices: no voltage applied to the gate allows maximum current through the source and drain.

Also take note that the amount of current allowed through a JFET is determined by a *voltage* signal rather than a *current* signal as with bipolar transistors. In fact, with the gate-source PN junction reverse-biased, there should be nearly zero current through the gate connection. For this reason, we classify the JFET as a *voltage-controlled device,* and the bipolar transistor as a *current-controlled device.*

If the gate-source PN junction is forward-biased with a small voltage, the JFET channel will "open" a little more to allow greater currents through. However, the PN junction of a JFET is not built to handle any substantial current itself, and thus it is not recommended to forward-bias the junction under any circumstances.

This is a very condensed overview of JFET operation. In the next section, we'll explore the use of the JFET as a switching device.

The Transistor as a Switch

Like its bipolar cousin, the field-effect transistor may be used as an on/off switch controlling electrical power to a load. Let's begin our investigation of the JFET as a switch with our familiar switch/lamp circuit:

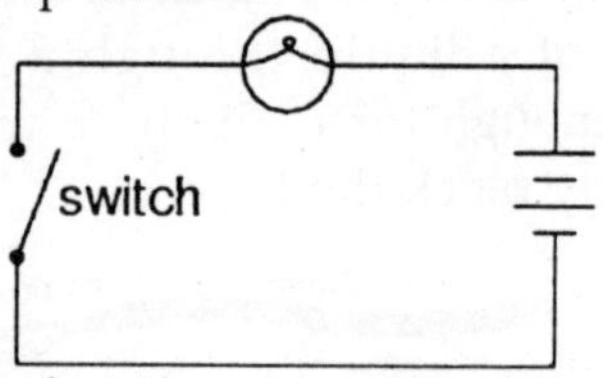

Remembering that the *controlled* current in a JFET flows between source and drain, we substitute the source and drain connections of a JFET for the two ends of the switch in the above circuit:

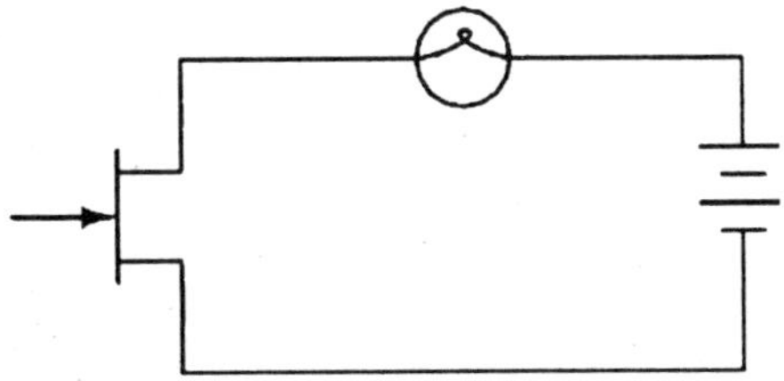

If you haven't noticed by now, the source and drain connections on a JFET look identical on the schematic symbol. Unlike the bipolar junction transistor where the emitter is clearly distinguished from the collector by the arrowhead, a JFET's source and drain lines both run perpendicular into the bar representing the semiconductor channel. This is no accident, as the source and drain lines of a JFET are often interchangeable in practice! In other words, JFETs are usually able to handle channel current in either direction, from source to drain or from drain to source.

Now all we need in the circuit is a way to control the JFET's conduction. With zero applied voltage between gate and source, the JFET's channel will be "open," allowing full current to the lamp. In order to turn the lamp off, we will need to connect another source of DC voltage between the gate and source connections of the JFET like this:

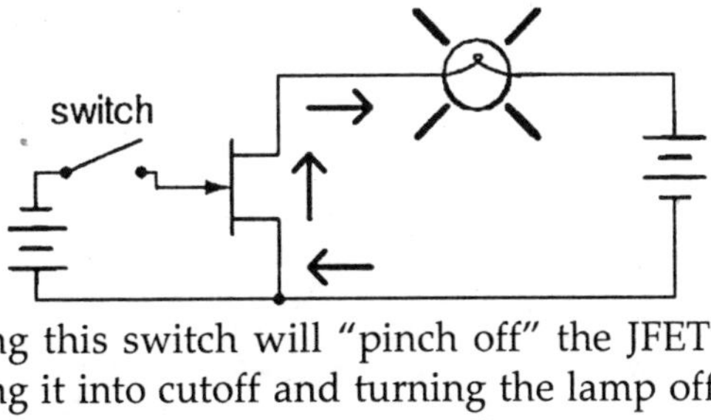

Closing this switch will "pinch off" the JFET's channel, thus forcing it into cutoff and turning the lamp off:

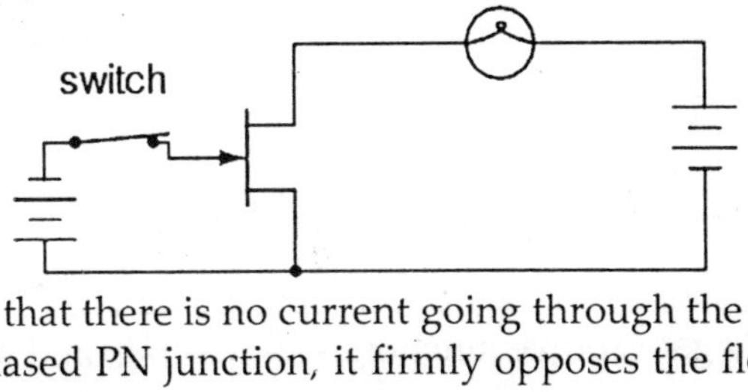

Note that there is no current going through the gate. As a reverse-biased PN junction, it firmly opposes the flow of any electrons through it. As a voltage-controlled device, the JFET requires negligible input current. This is an advantageous trait of the JFET over the bipolar transistor: there is virtually zero power required of the controlling signal.

Opening the control switch again should disconnect the reverse-biasing DC voltage from the gate, thus allowing the transistor to turn back on. Ideally, anyway, this is how it works. In practice this may not work at all:

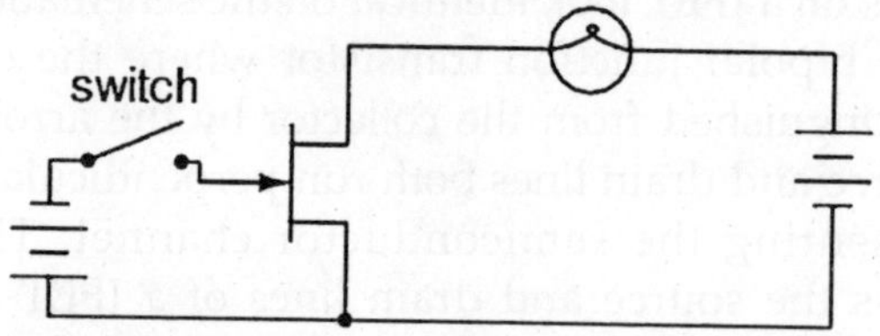

No lamp current after the switch opens!

Why is this? Why doesn't the JFET's channel open up again and allow lamp current through like it did before with no voltage applied between gate and source? The answer lies in the operation of the reverse-biased gate-source junction. The depletion region within that junction acts as an insulating barrier separating gate from source. As such, it possesses a certain amount of *capacitance* capable of storing an electric charge potential. After this junction has been forcibly reverse-biased by the application of an external voltage, it will tend to hold that reverse-biasing voltage as a stored charge even after the source of that voltage has been disconnected. What is needed to turn the JFET on again is to bleed off that stored charge between the gate and source through a resistor:

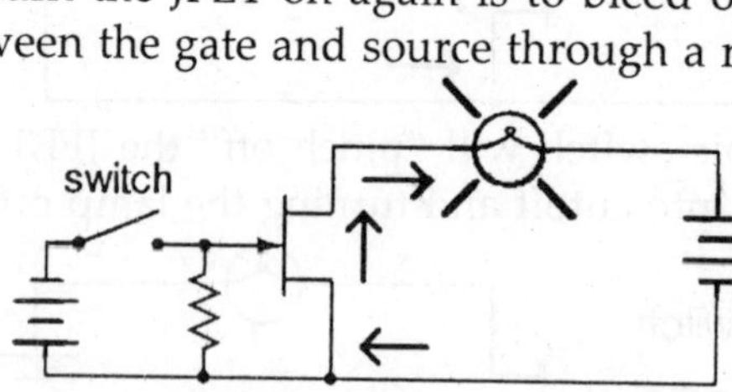

Resistor bleeds off stored charge in PN junction to allow transistor to turn on once again.

This resistor's value is not very important. The capacitance of the JFET's gate-source junction is very small, and so even a rather high-value bleed resistor creates a fast RC time constant, allowing the transistor to resume conduction with little delay once the switch is opened.

Like the bipolar transistor, it matters little where or what the controlling voltage comes from. We could use a solar cell, thermocouple, or any other sort of voltage-generating device to supply the voltage controlling the JFET's conduction. All that is required of a voltage source for JFET switch operation is *sufficient* voltage to achieve pinch-off of the JFET channel. This level is usually in the realm of a few volts DC, and is termed the *pinch-off* or *cutoff* voltage. The exact pinch-off voltage for any given JFET is a function of its unique design, and is not a universal figure like 0.7 volts is for a silicon BJT's base-emitter junction voltage.

METER CHECK OF A TRANSISTOR

Testing a JFET with a multimeter might seem to be a relatively easy task, seeing as how it has only one PN junction to test: either measured between gate and source, or between gate and drain.

Testing continuity through the drain-source channel is another matter, though. Remember from the last section how a stored charge across the capacitance of the gate-channel PN junction could hold the JFET in a pinched-off state without any external voltage being applied across it?

This can occur even when you're holding the JFET in your hand to test it! Consequently, any meter reading of continuity through that channel will be unpredictable, since you don't necessarily know if a charge is being stored by the gate-channel junction. Of course, if you know beforehand which terminals on the device are the gate, source, and drain, you may connect a jumper wire between gate and source to eliminate any stored charge and then proceed to test source-drain continuity with no problem. However, if you *don't* know which terminals are which, the unpredictability of the source-drain connection may confuse your determination of terminal identity.

A good strategy to follow when testing a JFET is to insert the pins of the transistor into anti-static foam (the material used to ship and store static-sensitive electronic components) just prior to testing. The conductivity of the foam will make a resistive connection between all terminals of the transistor when it is inserted. This connection will ensure that all residual voltage built up across the gate-channel PN junction will be neutralized, thus "opening up" the channel for an accurate meter test of source-to-drain continuity.

Since the JFET channel is a single, uninterrupted piece of semiconductor material, there is usually no difference between the source and drain terminals. A resistance check from source to drain should yield the same value as a check from drain to source. This resistance should be relatively low (a few hundred ohms at most) when the gate-source PN junction voltage is zero. By applying a reverse-bias voltage between gate and source, pinch-off of the channel should be apparent by an increased resistance reading on the meter.

This simulation directly reveals an important characteristic of the junction field-effect transistor: the control effect of gate voltage over drain current is *nonlinear*. Notice how the drain current does not decrease linearly as the gate-source voltage is increased. With the bipolar junction transistor, collector current was directly proportional to base current: output signal proportionately followed input signal. Not so with the JFET! The controlling signal (gate-source voltage) has less and less effect over the drain current as it approaches cutoff. In this simulation, most of the controlling action (75 percent of drain current decrease — from 400 μA to 100 μA) takes place within the first volt of gate-source voltage (from 0 to 1 volt), while the remaining 25 percent of drain current reduction takes another whole volt worth of input signal. Cutoff occurs at 2 volts input.

Linearity is generally important for a transistor because it allows it to faithfully amplify a waveform without distorting it. If a transistor is nonlinear in its input/output amplification, the shape of the input waveform will become corrupted in some way, leading to the production of harmonics in the output signal. The only time linearity is *not* important in a transistor circuit is when its being operated at the extreme limits of cutoff and saturation (off and on, respectively, like a switch).

A JFET's characteristic curves display the same current-regulating behavior as for a BJT, and the nonlinearity between gate-to-source voltage and drain current is evident in the disproportionate vertical spacings between the curves:

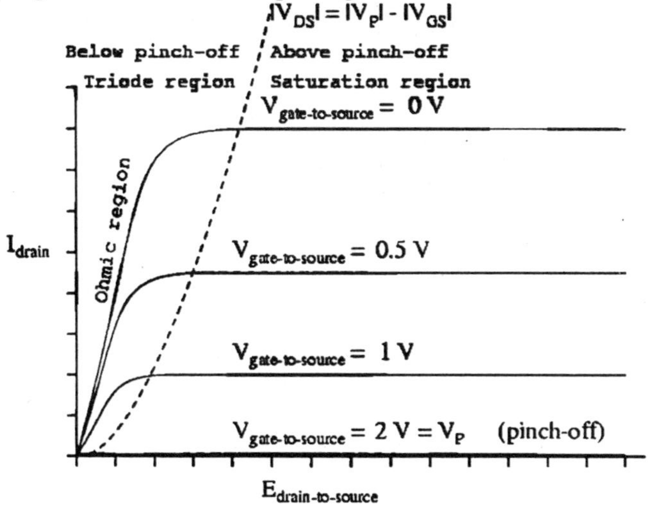

To better comprehend the current-regulating behavior of the JFET, it might be helpful to draw a model made up of simpler, more common components, just as we did for the BJT:

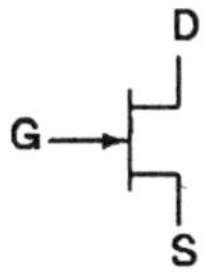

N-channel JFET diode-regulating diode model

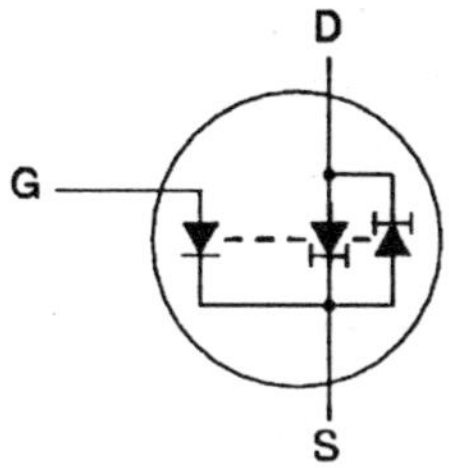

In the case of the JFET, it is the *voltage* across the reverse-biased gate-source diode which sets the current regulation point for the pair of constant-current diodes. A pair of opposing constant-current diodes is included in the model to facilitate current in either direction between source and drain, a trait made possible by the unipolar nature of the channel. With no PN junctions for the source-drain current to traverse, there is no polarity sensitivity in the controlled current. For this reason, JFETs are often referred to as *bilateral* devices.

A contrast of the JFET's characteristic curves against the curves for a bipolar transistor reveals a notable difference: the linear (straight) portion of each curve's non-horizontal area is surprisingly long compared to the respective portions of a BJT's characteristic curves:

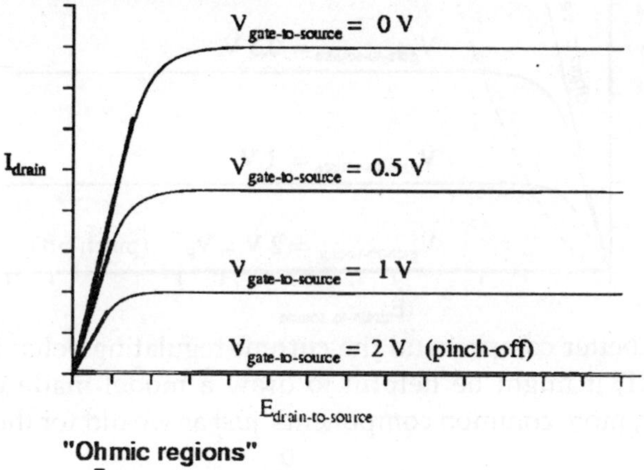

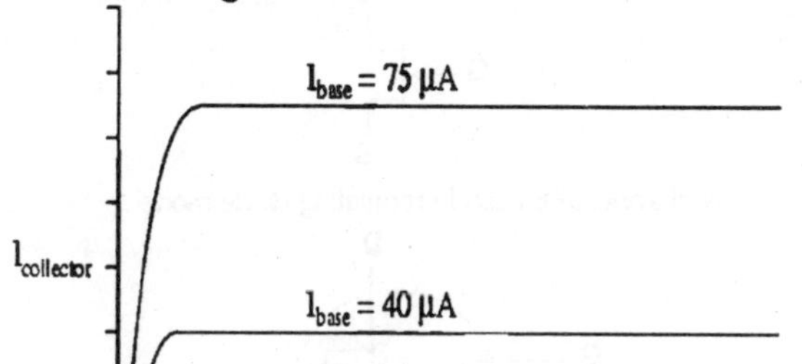

A JFET transistor operated in the *triode region* tends to act very much like a plain resistor as measured from drain to source. Like all simple resistances, its current/voltage graph is a straight line.

For this reason, the triode region (non-horizontal) portion of a JFET's characteristic curve is sometimes referred to as the *ohmic region*. In this mode of operation where there isn't enough drain-to-source voltage to bring drain current up to the regulated point, the drain current is directly proportional to the drain-to-source voltage. In a carefully designed circuit, this phenomenon can be used to an advantage. Operated in this region of the curve, the JFET acts like a voltage-controlled *resistance* rather than a voltage-controlled *current regulator*, and the appropriate model for the transistor is different:

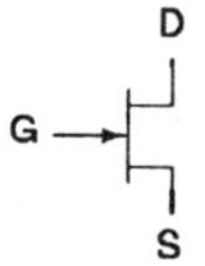

N-channel JFET diode-rheostat model
(for saturation, or "ohmic," mode only!)

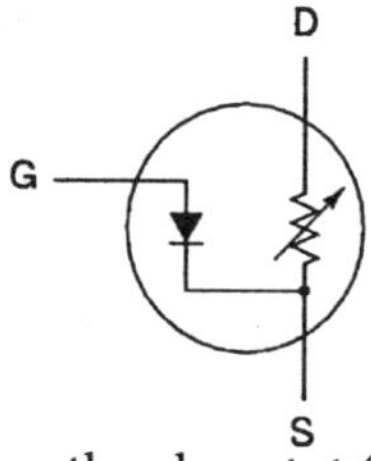

Here and here alone the rheostat (variable resistor) model of a transistor is accurate. It must be remembered, however, that this model of the transistor holds true only for a narrow range of its operation: when it is extremely saturated (far less voltage applied between drain and source than what is needed to achieve full regulated current through the drain). The amount of resistance (measured in ohms) between drain and source in this mode is controlled by how much reverse-bias voltage is applied between gate and source. The less gate-to-source voltage, the less resistance (steeper line on graph).

Because JFETs are *voltage*-controlled current regulators (at least when they're allowed to operate in their active), their inherent amplification factor cannot be expressed as a unitless ratio as with BJTs. In other words, there is no β ratio for a JFET. This is true for all voltage-controlled active devices, including other types of field-effect transistors and even electron tubes. There is, however, an expression of controlled (drain) current to controlling (gate-source) voltage, and it is called *transconductance*. Its unit is Siemens, the same unit for conductance (formerly known as the *mho*).

Why this choice of units? Because the equation takes on the general form of current (output signal) divided by voltage (input signal).

$$g_{fs} = \frac{\Delta I_D}{\Delta V_{GS}}.$$

Where,

g_{fs} = Transconductance in Siemens

ΔI_D= Change indrain current

ΔV_{GS} = Change in gate-source voltage.

Unfortunately, the transconductance value for any JFET is not a stable quantity: it varies significantly with the amount of gate-to-source control voltage applied to the transistor. The drain current does not change proportionally with changes in gate-source voltage.

To calculate drain current for any given gate-source voltage, there is another equation that may be used. It is obviously nonlinear upon inspection (note the power of 2), reflecting the nonlinear behavior we've already experienced in simulation:

$$I_D = I_{DSS}\left(1 - \frac{V_{GS}}{V_{GS(cutoff)}}\right)^2.$$

Where,

I_D = Drain current

I_{DSS} = Drain current with gate shorted to source

V_{GS} = Gate-to-source voltage

$V_{GS(cutoff)}$ = Pinch-off gate-to-source voltage.

INSULATED-GATE FIELD-EFFECT TRANSISTORS

There is more than one type of field-effect transistor, the junction field-effect transistor, or JFET, uses voltage applied across a reverse-biased PN junction to control the width of that junction's depletion region, which then controls the conductivity of a semiconductor channel through which the controlled current moves.

Another type of field-effect device — the insulated gate field-effect transistor, or IGFET — exploits a similar principle of a depletion region controlling conductivity through a semiconductor channel, but it differs primarily from the JFET in that there is no *direct* connection between the gate lead and the semiconductor material itself. Rather, the gate lead is insulated from the transistor body by a thin barrier, hence the term *insulated gate.*

This insulating barrier acts like the dielectric layer of a capacitor, and allows gate-to-source voltage to influence the depletion region electrostatically rather than by direct connection.

In addition to a choice of N-channel versus P-channel design, IGFETs come in two major types: *enhancement* and *depletion.* The depletion type is more closely related to the JFET, so we will begin our study of IGFETs with it.

DEPLETION-TYPE IGFETS

Insulated gate field-effect transistors are unipolar devices just like JFETs: that is, the controlled current does not have to cross a PN junction. There is a PN junction inside the transistor, but its only purpose is to provide that nonconducting depletion region which is used to restrict current through the channel.

Here is a diagram of an N-channel IGFET of the "depletion" type:

Notice how the source and drain leads connect to either end of the N channel, and how the gate lead attaches to a metal plate separated from the channel by a thin insulating barrier. That barrier is sometimes made from silicon dioxide (the primary chemical compound found in sand), which is a very good insulator.

N-channel, D-type **IGFET**

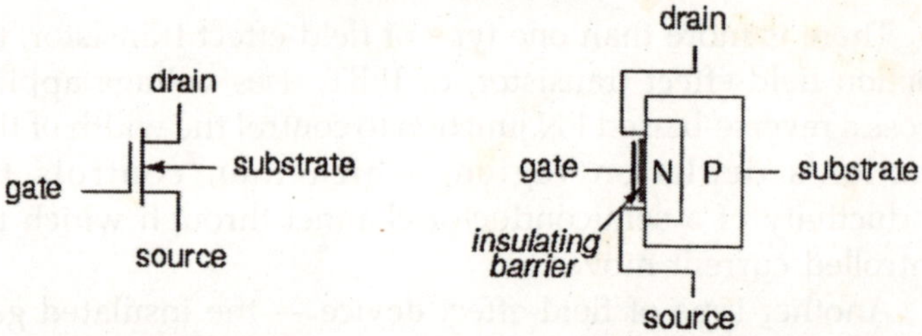

schematic symbol ***physical diagram***

Due to this Metal (gate) - Oxide (barrier) - Semiconductor (channel) construction, the IGFET is sometimes referred to as a MOSFET. There are other types of IGFET construction, though, and so "IGFET" is the better descriptor for this general class of transistors.

Notice also how there are four connections to the IGFET. In practice, the *substrate* lead is directly connected to the *source* lead to make the two electrically common. Usually, this connection is made internally to the IGFET, eliminating the separate substrate connection, resulting in a three-terminal device with a slightly different schematic symbol:

N-channel, D-type **IGFET**

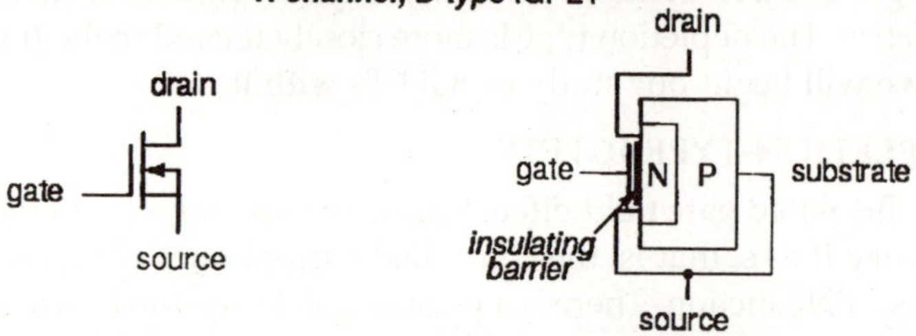

schematic symbol ***physical diagram***

With source and substrate common to each other, the N and P layers of the IGFET end up being directly connected to each other through the outside wire. This connection prevents any voltage from being impressed across the PN junction. As a result, a depletion region exists between the two materials, but it can never be expanded or collapsed. JFET operation is based on the expansion of the PN junction's depletion region, but here in the IGFET that cannot happen, so IGFET operation must be based on a different effect.

Indeed it is, for when a controlling voltage is applied between gate and source, the conductivity of the channel is changed as a result of the depletion region *moving* closer to or further away from the gate.

In other words, the channel's effective width changes just as with the JFET, but this change in channel width is due to depletion region *displacement* rather than depletion region *expansion*.

In an N-channel IGFET, a controlling voltage applied positive (+) to the gate and negative (-) to the source has the effect of repelling the PN junction's depletion region, expanding the N-type channel and increasing conductivity:

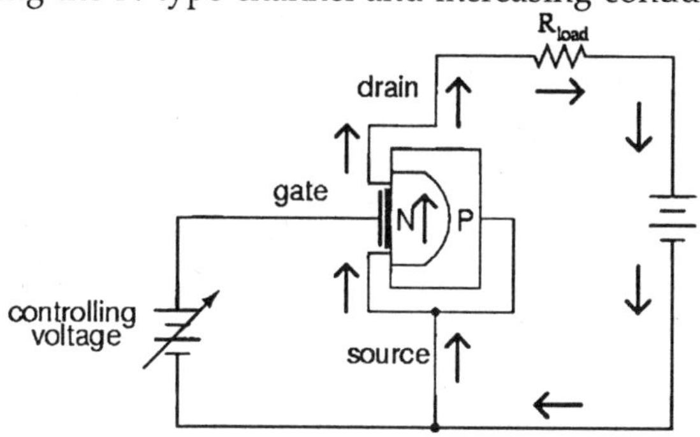

Channel expands for greater conductivity

Reversing the controlling voltage's polarity has the opposite effect, attracting the depletion region and narrowing the channel, consequently reducing channel conductivity:

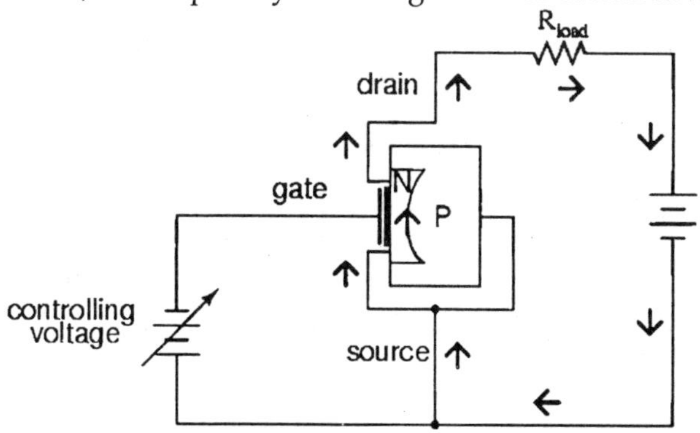

Channel narrows for less conductivity

The insulated gate allows for controlling voltages of any polarity without danger of forward-biasing a junction, as was the concern with JFETs. This type of IGFET, although its called a "depletion-type," actually has the capability of having its channel *either* depleted (channel narrowed) *or* enhanced (channel expanded). Input voltage polarity determines which way the channel will be influenced.

Understanding which polarity has which effect is not as difficult as it may seem. The key is to consider the type of semiconductor doping used in the channel (N-channel or P-channel?), then relate that doping type to the side of the input voltage source connected to the channel by means of the source lead. If the IGFET is an N-channel and the input voltage is connected so that the positive (+) side is on the gate while the negative (–) side is on the source, the channel will be enhanced as extra electrons build up on the channel side of the dielectric barrier. Think, "negative (–) correlates with *N*-type, thus enhancing the channel with the right type of charge carrier (electrons) and making it more conductive." Conversely, if the input voltage is connected to an N-channel IGFET the other way, so that negative (-) connects to the gate while positive (+) connects to the source, free electrons will be "robbed" from the channel as the gate-channel capacitor charges, thus depleting the channel of majority charge carriers and making it less conductive.

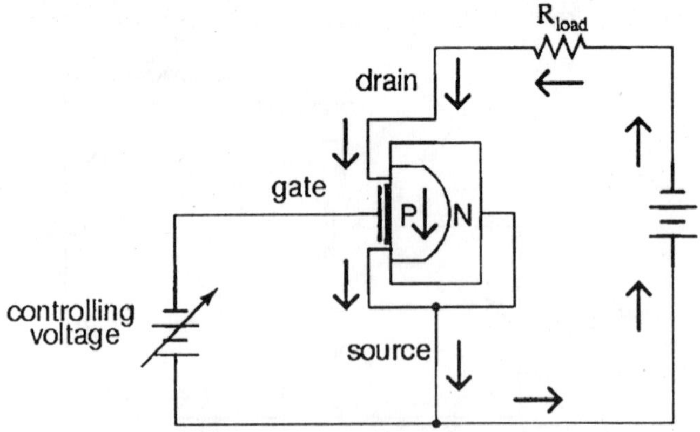

Channel expands for greater conductivity

For P-channel IGFETs, the input voltage polarity and channel effects follow the same rule. That is to say, it takes just the opposite polarity as an N-channel IGFET to either deplete or enhance:

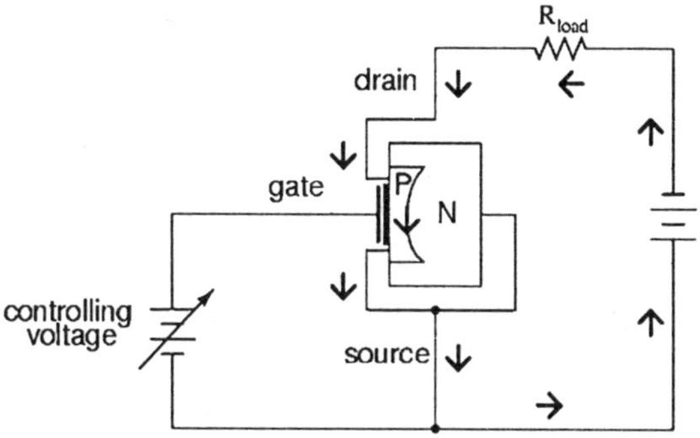

Channel narrows for less conductivity

Illustrating the proper biasing polarities with standard IGFET symbols:

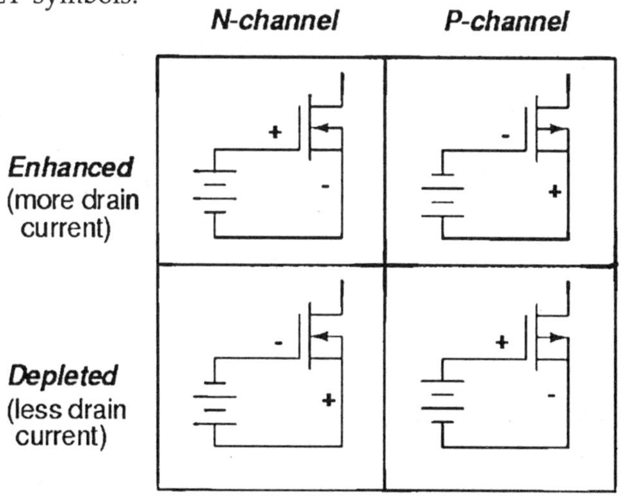

When there is zero voltage applied between gate and source, the IGFET will conduct current between source and drain, but not as much current as it would if it were enhanced by the proper gate voltage. This places the depletion-type, or simply *D-type,* IGFET in a category of its own in the transistor world. Bipolar junction transistors are *normally-off* devices: with no base current, they block any current from going through the collector. Junction field-effect transistors are *normally-on* devices: with zero applied gate-to-source voltage, they allow maximum drain current (actually, you can coax a JFET into greater drain currents by applying a very small forward-bias voltage between gate and source, but this should never be done in practice for risk of damaging its fragile PN junction).

D-type IGFETs, however, are *normally half-on* devices: with no gate-to-source voltage, their conduction level is somewhere between cutoff and full saturation. Also, they will tolerate applied gate-source voltages of any polarity, the PN junction being immune from damage due to the insulating barrier and especially the direct connection between source and substrate preventing any voltage differential across the junction.

Ironically, the conduction behavior of a D-type IGFET is strikingly similar to that of an electron tube of the triode/tetrode/pentode variety. These devices were voltage-controlled current regulators that likewise allowed current through them with zero controlling voltage applied. A controlling voltage of one polarity (grid negative and cathode positive) would diminish conductivity through the tube while a voltage of the other polarity (grid positive and cathode negative) would enhance conductivity. I find it curious that one of the later transistor designs invented exhibits the same basic properties of the very first active (electronic) device.

IGBTS

Because of their insulated gates, IGFETs of all types have extremely high current gain: there can be no sustained gate current if there is no continuous gate *circuit* in which electrons may continually flow. The only current we see through the gate terminal of an IGFET, then, is whatever transient (brief surge) may be required to charge the gate-channel capacitance and displace the depletion region as the transistor switches from an "on" state to an "off" state, or vice versa.

This high current gain would at first seem to place IGFET technology at a decided advantage over bipolar transistors for the control of very large currents. If a bipolar junction transistor is used to control a large collector current, there must be a substantial base current sourced or sunk by some control circuitry, in accordance with the β ratio. To give an example, in order for a power BJT with a β of 20 to conduct a collector current of 100 amps, there must be at least 5 amps of base current, a substantial amount of current in itself for miniature discrete or integrated control circuitry to handle:

It would be nice from the standpoint of control circuitry to have power transistors with high current gain, so that far less current is needed for control of load current. Of course, we can use Darlington pair transistors to increase the current gain, but this kind of arrangement still requires *far* more controlling current than an equivalent power IGFET:

Unfortunately, though, IGFETs have problems of their own controlling high current: they typically exhibit greater drain-to-source voltage drop while saturated than the collector-to-emitter voltage drop of a saturated BJT. This greater voltage drop equates to higher power dissipation for the same amount of load current, limiting the usefulness of IGFETs as high-power devices. Although some specialized designs such as the so-called VMOS transistor have been designed to minimize this inherent disadvantage, the bipolar junction transistor is still superior in its ability to switch high currents.

An interesting solution to this dilemma leverages the best features of IGFETs with the best of features of BJTs, in one device called an *Insulated-Gate Bipolar Transistor*, or *IGBT*. Also known as an *Bipolar-mode MOSFET*, a *Conductivity-Modulated Field-Effect Transistor* (*COMFET*), or simply as an *Insulated-Gate Transistor* (*IGT*), it is equivalent to a Darlington pair of IGFET and BJT:

In essence, the IGFET controls the base current of a BJT, which handles the main load current between collector and emitter. This way, there is extremely high current gain (since the insulated gate of the IGFET draws practically no current from the control circuitry), but the collector-to-emitter voltage drop during full conduction is as low as that of an ordinary BJT.

One disadvantage of the IGBT over a standard BJT is its slower turn-off time. For *fast* switching and high current-handling capacity, its difficult to beat the bipolar junction transistor. Faster turn-off times for the IGBT may be achieved by certain changes in design, but only at the expense of a higher saturated voltage drop between collector and emitter. However, the IGBT provides a good alternative to IGFETs and BJTs for high-power control applications.

Chapter 5

Thyristors

HYSTERESIS

Thyristors are a class of semiconductor components exhibiting *hysteresis,* that property whereby a system fails to return to its original state after some cause of state change has been removed. A very simple example of hysteresis is the mechanical action of a toggle switch: when the lever is pushed, it flips to one of two extreme states (positions) and will remain there even after the source of motion is removed (after you remove your hand from the switch lever). To illustrate the absence of hysteresis, consider the action of a "momentary" pushbutton switch, which returns to its original state after the button is no longer pressed: when the stimulus is removed (your hand), the system (switch) immediately and fully returns to its prior state with no "latching" behavior.

Bipolar, junction field-effect, and insulated gate field-effect transistors are all non-hysteretic devices. That is, they do not inherently "latch" into a state after being stimulated by a voltage or current signal. For any given input signal at any given time, a transistor will exhibit a predictable output response as defined by its characteristic curve. Thyristors, on the other hand, are semiconductor devices that tend to stay "on" once turned on, and tend to stay "off" once turned off. A momentary event is able to flip these devices into either their on or off states where they will remain that way on their own, even after the cause of the state change is taken away. As such, they are useful only as on/off switching devices — much like a toggle switch — and cannot be used as analog signal amplifiers.

Thyristors are constructed using the same technology as bipolar junction transistors, and in fact may be analyzed as circuits comprised of transistor pairs. How then, can a hysteretic device (a thyristor) be made from non-hysteretic devices (transistors)? The answer to this question is *positive feedback,* also known as *regenerative feedback.* As you should recall, feedback is the condition where a percentage of the output signal is "fed back" to the input of an amplifying device. Negative, or degenerative, feedback results in a diminishing of voltage gain with increases in stability, linearity, and bandwidth. Positive feedback, on the other hand, results in a kind of instability where the amplifier's output tends to "saturate." In the case of thyristors, this saturating tendency equates to the device "wanting" to stay on once turned on, and off once turned off.

GAS DISCHARGE TUBES

If you've ever witnessed a lightning storm, you've seen electrical hysteresis in action (and probably didn't realize what you were seeing). The action of strong wind and rain accumulates tremendous static electric charges between cloud and earth, and between clouds as well. Electric charge imbalances manifest themselves as high voltages, and when the electrical resistance of air can no longer hold these high voltages at bay, huge surges of current travel between opposing poles of electrical charge which we call "lightning."

The buildup of high voltages by wind and rain is a fairly continuous process, the rate of charge accumulation increasing under the proper atmospheric conditions. However, lightning bolts are anything but continuous: they exist as relatively brief surges rather than continuous discharges. Why is this? Why don't we see soft, glowing lightning *arcs* instead of violently brief lightning *bolts*? The answer lies in the nonlinear (and hysteretic) resistance of air.

Under ordinary conditions, air has an extremely high amount of resistance. It is so high, in fact, that we typically treat its resistance as infinite and electrical conduction through the air as negligible. The presence of water and/or dust in air lowers its resistance some, but it is still an insulator for most practical purposes. When a sufficient amount of high voltage is applied across a distance of air, though, its electrical properties change: electrons become "stripped" from their normal positions around their respective atoms and are liberated to constitute a current. In this state, air is considered to be *ionized* and is referred to as a *plasma* rather than a normal *gas*. This usage of the word "plasma" is not to be confused with the medical term (meaning the fluid portion of blood), but is a fourth state of matter, the other three being solid, liquid, and vapor (gas). Plasma is a relatively good conductor of electricity, its specific resistance being much lower than that of the same substance in its gaseous state.

As an electric current moves through the plasma, there is energy dissipated in the plasma in the form of heat, just as current through a solid resistor dissipates energy in the form of heat. In the case of lightning, the temperatures involved are extremely high. High temperatures are also sufficient to convert gaseous air into a plasma or maintain plasma in that state without the presence of high voltage. As the voltage between cloud and earth, or between cloud and cloud, decreases as the charge imbalance is neutralized by the current of the lightning bolt, the heat dissipated by the bolt maintains the air path in a plasma state, keeping its resistance low. The lightning bolt remains a plasma until the voltage decreases to too low a level to sustain enough current to dissipate enough heat. Finally, the air returns to a normal, gaseous state and stops conducting current, thus allowing voltage to build up once more.

Note how throughout this cycle, the air exhibits hysteresis. When not conducting electricity, it tends to *remain an insulator* until voltage builds up past a critical threshold point. Then, once it changes state and becomes a plasma, it tends to *remain a conductor* until voltage falls below a lower critical threshold point. Once "turned on" it tends to stay "on," and once "turned off" it tends to stay "off." This hysteresis, combined with a steady buildup of voltage due to the electrostatic effects of wind and rain, explains the action of lightning as brief bursts.

In electronic terms, what we have here in the action of lightning is a simple *relaxation oscillator*. Oscillators are electronic circuits that produce an oscillating (AC) voltage from a steady supply of DC power.

A relaxation oscillator is one that works on the principle of a charging capacitor that is suddenly discharged every time its voltage reaches a critical threshold value. One of the simplest relaxation oscillators in existence is comprised of three components (not counting the DC power supply): a resistor, capacitor, and neon lamp:

Simple relaxation oscillator

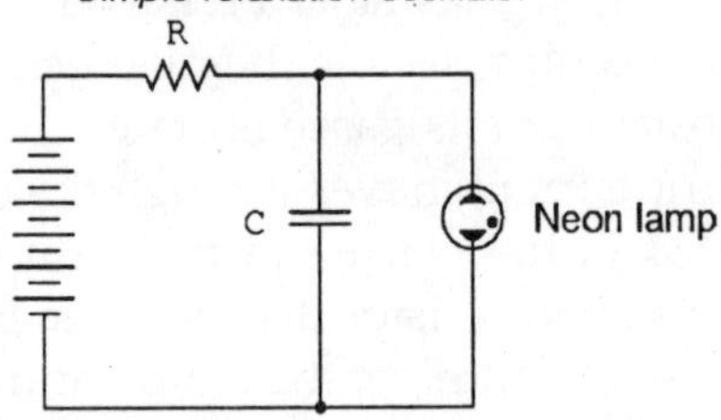

Neon lamps are nothing more than two metal electrodes inside a sealed glass bulb, separated by the neon gas inside. At room temperatures and with no applied voltage, the lamp has nearly infinite resistance. However, once a certain threshold voltage is exceeded (this voltage depends on the gas pressure and geometry of the lamp), the neon gas will become ionized (turned into a plasma) and its resistance dramatically reduced. In effect, the neon lamp exhibits the same characteristics as air in a lightning storm, complete with the emission of light as a result of the discharge, albeit on a much smaller scale.

The capacitor in the relaxation oscillator circuit shown above charges at an inverse exponential rate determined by the size of the resistor. When its voltage reaches the threshold voltage of the lamp, the lamp suddenly "turns on" and quickly discharges the capacitor to a low voltage value. Once discharged, the lamp "turns off" and allows the capacitor to build up a charge once more. The result is a series of brief flashes of light from the lamp, the rate of which dictated by battery voltage, resistor resistance, capacitor capacitance, and lamp threshold voltage.

While gas-discharge lamps are more commonly used as sources of illumination, their hysteretic properties were leveraged in slightly more sophisticated variants known as *thyratron tubes*. Essentially a gas-filled triode tube (a triode being a three-element vacuum electron tube performing much a similar function to the N-channel, D-type IGFET), the thyratron tube could be turned on with a small control voltage applied between grid and cathode, and turned off by reducing the plate-to-cathode voltage.

(Simple) Thyratron control circuit

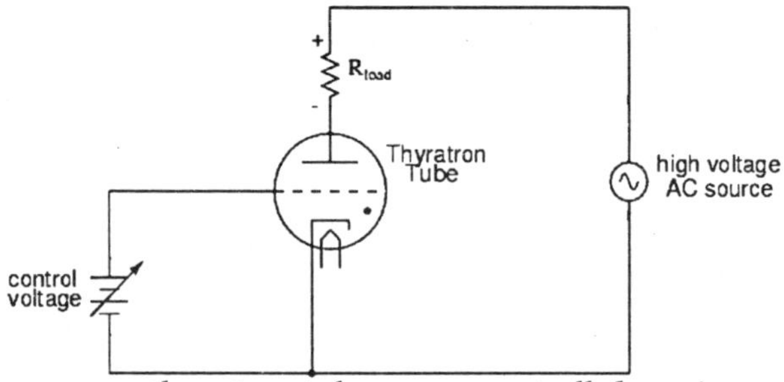

In essence, thyratron tubes were *controlled* versions of neon lamps built specifically for switching current to a load. The dot inside the circle of the schematic symbol indicates a gas fill, as opposed to the hard vacuum normally seen in other electron tube designs. In the circuit shown above, the thyratron tube allows current through the load in one direction (note the polarity across the load resistor) when triggered by the small DC control voltage connected between grid and cathode. Note that the load's power source is AC, which provides a clue as to how the thyratron turns off after its been triggered on: since AC voltage periodically passes through a condition of 0 volts between half-cycles, the current through an AC-powered load must also periodically halt. This brief pause of current between half-cycles gives the tube's gas time to cool, letting it return to its normal "off" state. Conduction may resume only if there is enough voltage applied by the AC power source (some other time in the wave's cycle) *and* if the DC control voltage allows it.

An oscilloscope display of load voltage in such a circuit would look something like this:

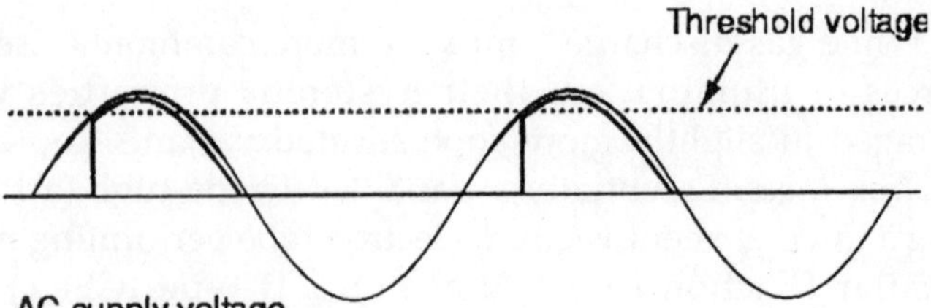

AC supply voltage
Load voltage

As the AC supply voltage climbs from zero volts to its first peak, the load voltage remains at zero (no load current) until the threshold voltage is reached. At that point, the tube switches "on" and begins to conduct, the load voltage now following the AC voltage through the rest of the half cycle. Notice how there is load voltage (and thus load current) even when the AC voltage waveform has dropped below the threshold value of the tube.

This is hysteresis at work: the tube stays in its conductive mode past the point where it first turned on, continuing to conduct until there the supply voltage drops off to almost zero volts. Because thyratron tubes are one-way (diode) devices, there is no voltage across the load through the negative half-cycle of AC. In practical thyratron circuits, multiple tubes arranged in some form of full-wave rectifier circuit to facilitate full-wave DC power to the load.

The thyratron tube has been applied to a relaxation oscillator circuit. The frequency is controlled by a small DC voltage between grid and cathode. This voltage-controlled oscillator is known as a *VCO*. Relaxation oscillators produce a very non-sinusoidal output, and so they exist mostly as demonstration circuits (as is the case here) or in applications where the harmonic rich waveform is desirable.

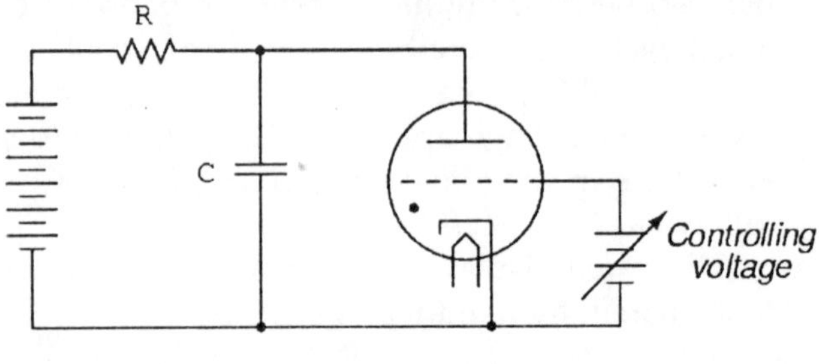

I speak of thyratron tubes in the past tense for good reason: modern semiconductor components have obsoleted thyratron tube technology for all but a few very special applications. It is no coincidence that the word *thyristor* bears so much similarity to the word *thyratron,* for this class of semiconductor components does much the same thing: use *hy*ster*e*tically switch current on and off. It is these modern devices that we now turn our attention to.

THE SHOCKLEY DIODE

Our exploration of thyristors begins with a device called the *four-layer diode,* also known as a *PNPN diode,* or a *Shockley diode* after its inventor, William Shockley. This is not to be confused with a *Schottky diode,* that two-layer metal-semiconductor device known for its high switching speed. A crude illustration of the Shockley diode, often seen in textbooks, is a four-layer sandwich of P-N-P-N semiconductor material:

Shockley, or *4-layer*, diode

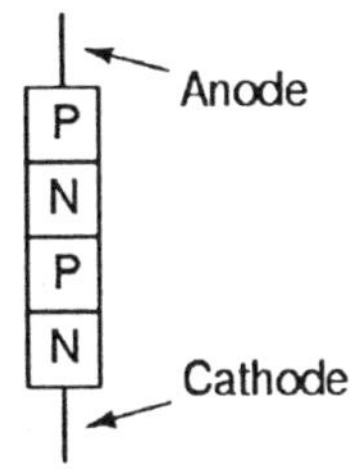

Unfortunately, this simple illustration does nothing to enlighten the viewer on how it works or why. Consider an alternative rendering of the device's construction:

Shockley, or *4-layer*, diode

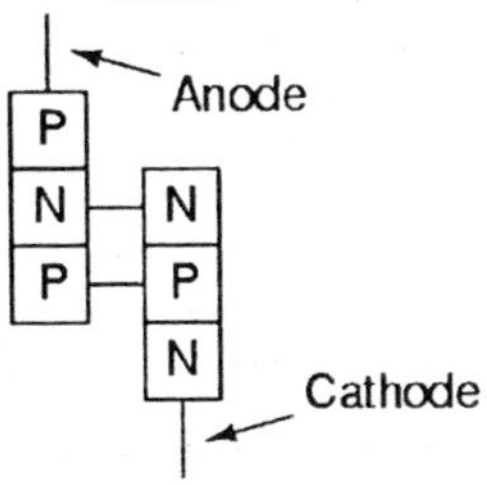

Shown like this, it appears to be a set of interconnected bipolar transistors, one PNP and the other NPN. Drawn using standard schematic symbols, and respecting the layer doping concentrations not shown in the last image, the Shockley diode looks like this:

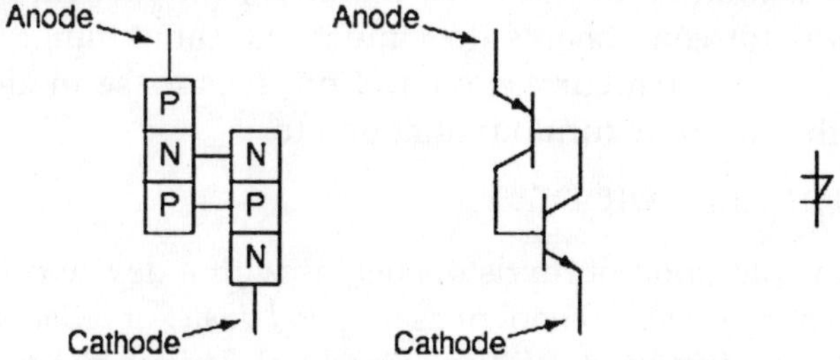

Physical diagram *Equivalent schematic* *Schematic symbol*

Let's connect one of these devices to a source of variable voltage and see what happens:

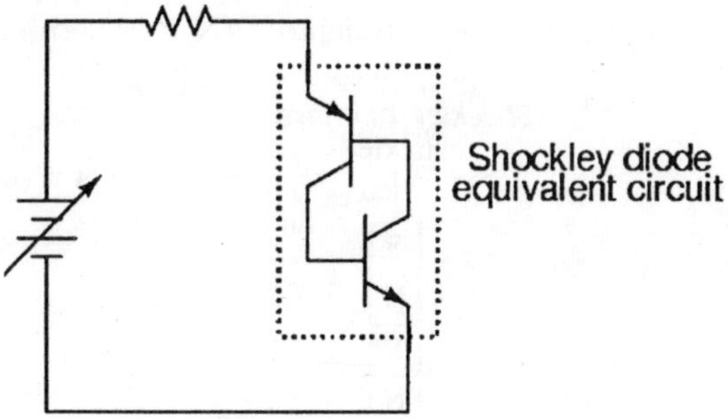

With no voltage applied, of course there will be no current. As voltage is initially increased, there will still be no current because neither transistor is able to turn on: both will be in cutoff mode. To understand why this is, consider what it takes to turn a bipolar junction transistor on: current through the base-emitter junction. As you can see in the diagram, base current through the lower transistor is controlled by the upper transistor, and the base current through the upper transistor is controlled by the lower transistor. In other words, neither transistor can turn on until the *other* transistor turns on. What we have here, in vernacular terms, is known as a Catch-22.

So how can a Shockley diode ever conduct current, if its constituent transistors stubbornly maintain themselves in a state of cutoff? The answer lies in the behavior of *real* transistors as opposed to *ideal* transistors. An ideal bipolar transistor will never conduct collector current if there is no base current, no matter how much or little voltage we apply between collector and emitter. Real transistors, on the other hand, have definite limits to how much collector-emitter voltage they can withstand before they break down and conduct.

If two real transistors are connected together in this fashion to form a Shockley diode, they *will* be able to conduct if there is sufficient voltage applied by the battery between anode and cathode to cause one of them to break down. Once one transistor breaks down and begins to conduct, it will allow base current through the other transistor, causing it to turn on in a normal fashion, which then allows base current through the first transistor.

The end result is that both transistors will be saturated, now keeping each other turned on instead of off. So, we can force a Shockley diode to turn on by applying sufficient voltage between anode and cathode. As we have seen, this will inevitably cause one of the transistors to turn on, which then turns the other transistor on, ultimately "latching" both transistors on where they will tend to remain. But how do we now get the two transistors to turn off again? Even if the applied voltage is reduced to a point well below what it took to get the Shockley diode conducting, it will remain conducting because both transistors now have base current to maintain regular, controlled conduction.

The answer to this is to reduce the applied voltage to a much lower point where there is too little current to maintain transistor bias, at which point one of the transistors will cutoff, which then halts base current through the other transistor, sealing both transistors in the "off" state as they were before any voltage was applied at all.

If we graph this sequence of events and plot the results on an I/V graph, the hysteresis is very evident. First, we will observe the circuit as the DC voltage source (battery) is set to zero voltage:

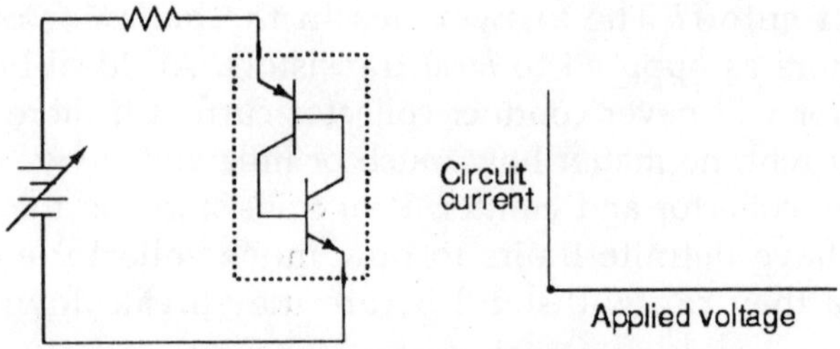

Zero applied voltage; zero current

Next, we will steadily increase the DC voltage. Current through the circuit is at or nearly at zero, as the breakdown limit has not been reached for either transistor:

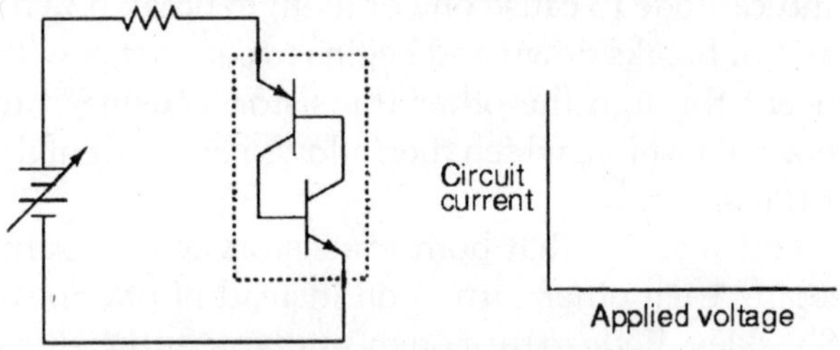

Some voltage applied, still no appreciable current

When the voltage breakdown limit of one transistor is reached, it will begin to conduct collector current even though no base current has gone through it yet. Normally, this sort of treatment would destroy a bipolar junction transistor, but the PNP junctions comprising a Shockley diode are engineered to take this kind of abuse, similar to the way a Zener diode is built to handle reverse breakdown without sustaining damage. For the sake of illustration I'll assume the lower transistor breaks down first, sending current through the base of the upper transistor:

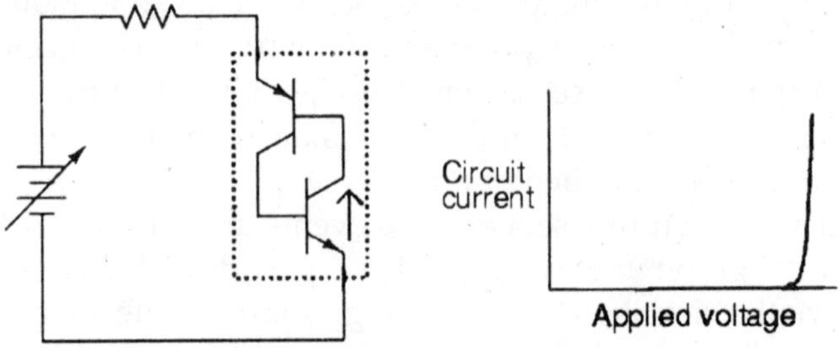

More voltage applied; lower transistor breaks down

As the upper transistor receives base current, it turns on as expected. This action allows the lower transistor to conduct normally, the two transistors "sealing" themselves in the "on" state. Full current is very quickly seen in the circuit:

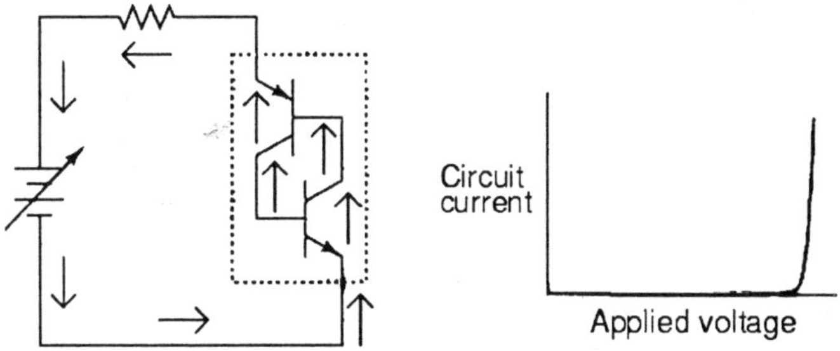

Transistors now fully conducting

When one transistor breaks down, it allows current through the device structure. This current may be viewed as the "output" signal of the device. Once an output current is established, it works to hold both transistors in saturation, thus ensuring the continuation of a substantial output current. In other words, an output current "feeds back" positively to the input (transistor base current) to keep both transistors in the "on" state, thus reinforcing (or *regenerating*) itself.

With both transistors maintained in a state of saturation with the presence of ample base current, they will continue to conduct even if the applied voltage is greatly reduced from the breakdown level. The effect of positive feedback is to keep both transistors in a state of saturation despite the loss of input stimulus (the original, high voltage needed to break down one transistor and cause a base current through the other transistor):

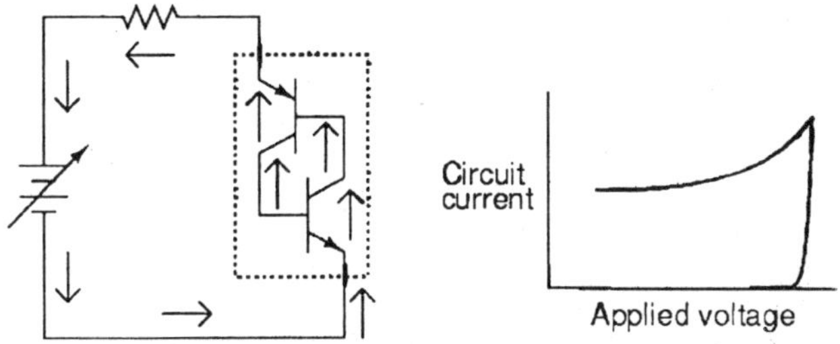

Current maintained even when voltage is reduced

If the DC voltage source is turned down too far, though, the circuit will eventually reach a point where there isn't enough current to sustain both transistors in saturation. As one transistor passes less and less collector current, it reduces the base current for the other transistor, thus reducing base current for the first transistor. The vicious cycle continues rapidly until both transistors fall into cutoff:

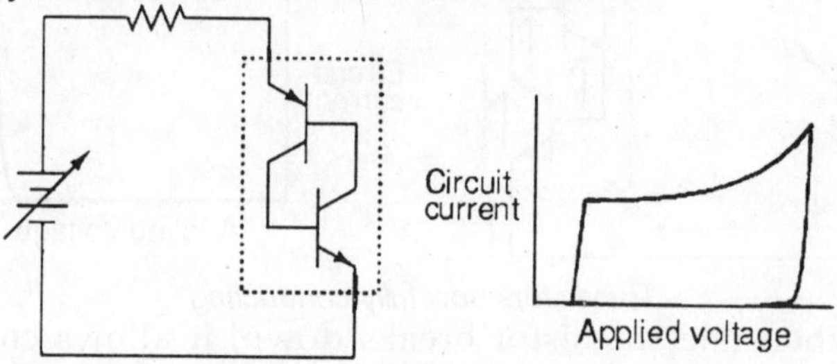

If the voltage drops too low, both transistors shut off

Here, positive feedback is again at work: the fact that the cause/effect cycle between both transistors is "vicious" (a decrease in current through one works to decrease current through the other, further decreasing current through the first transistor) indicates a positive relationship between output (controlled current) and input (controlling current through the transistors' bases).

The resulting curve on the graph is classicly hysteretic: as the input signal (voltage) is increased and decreased, the output (current) does not follow the same path going down as it did going up:

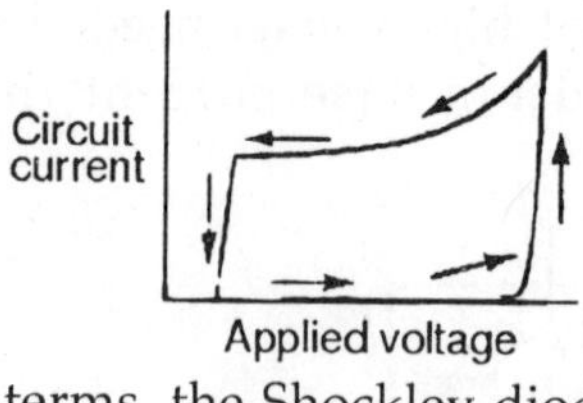

Put in simple terms, the Shockley diode tends to stay on once its turned on, and stay off once its turned off. There is no "in-between" or "active" mode in its operation: it is a purely on or off device, as are all thyristors.

There are a few special terms applied to Shockley diodes and all other thyristor devices built upon the Shockley diode foundation. First is the term used to describe its "on" state: *latched*. The word "latch" is reminiscent of a door lock mechanism, which tends to keep the door closed once it has been pushed shut. The term *firing* refers to the initiation of a latched state. In order to get a Shockley diode to latch, the applied voltage must be increased until *breakover* is attained. Despite the fact that this action is best described in terms of transistor break*down*, the term break*over* is used instead because the end result is a pair of transistors in mutual saturation rather than destruction as would be the case with a normal transistor. A latched Shockley diode is re-set back into its non conducting state by reducing current through it until *low-current dropout* occurs.

It should be noted that Shockley diodes may be fired in a way other than break over: excessive *voltage rise*, or *dv/dt*. This is when the applied voltage across the diode increases at a high rate of change. This is able to cause latching (turning on) of the diode due to inherent junction capacitances within the transistors. Capacitors, as you may recall, oppose *changes* in voltage by drawing or supplying current. If the applied voltage across a Shockley diode rises at too fast a rate, those tiny capacitances will draw enough current during that time to activate the transistor pair, turning them both on. Usually, this form of latching is undesirable, and can be minimized by filtering high-frequency (fast voltage rises) from the diode with series inductors and/or parallel resistor-capacitor networks called *snubbers*:

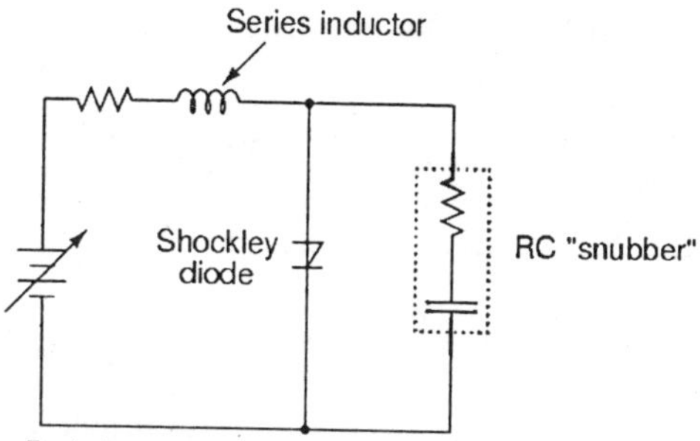

Both the series inductor and the parallel resistor-capacitor "snubber" circuit help minimize the Shockley diode's exposure to excessively rising voltages.

The voltage rise limit of a Shockley diode is referred to as the *critical rate of voltage rise*. Manufacturers usually provide this specification for the devices they sell.

THE DIAC

Like all diodes, Shockley diodes are unidirectional devices; that is, they only conduct current in one direction. If bidirectional (AC) operation is desired, two Shockley diodes may be joined in parallel facing different directions to form a new kind of thyristor, the *DIAC*:

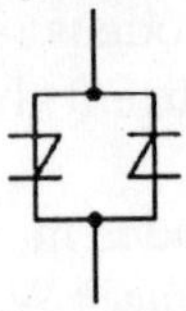

DIAC equivalent circuit DIAC schematic symbol

A DIAC operated with a DC voltage across it behaves exactly the same as a Shockley diode. With AC, however, the behavior is different from what one might expect. Because alternating current repeatedly reverses direction, DIACs will not stay latched longer than one-half cycle. If a DIAC becomes latched.

It will continue to conduct current only as long as there is voltage available to push enough current in that direction. When the AC polarity reverses, as it must twice per cycle, the DIAC will drop out due to insufficient current, necessitating another breakover before it conducts again. The result is a current waveform that looks like this:

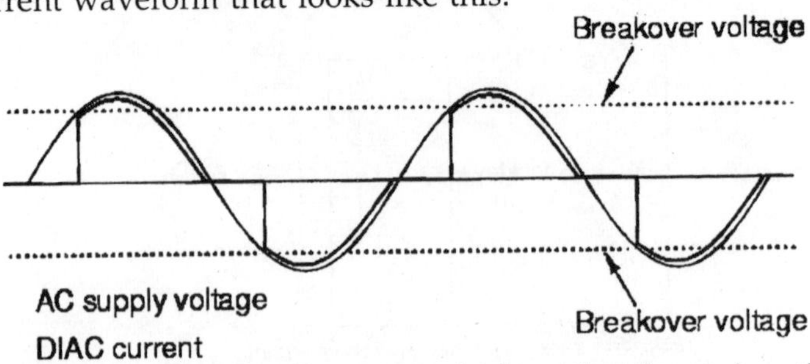

DIACs are almost never used alone, but in conjunction with other thyristor devices.

THE SILICON-CONTROLLED RECTIFIER (SCR)

Shockley diodes are curious devices, but rather limited in application. Their usefulness may be expanded, however, by equipping them with another means of latching. In doing so, they become true amplifying devices (if only in an on/off mode), and we refer to them as *silicon-controlled rectifiers*, or *SCR*s.

The progression from Shockley diode to SCR is achieved with one small addition, actually nothing more than a third wire connection to the existing PNPN structure:

The Silicon-Controlled Rectifier (SCR)

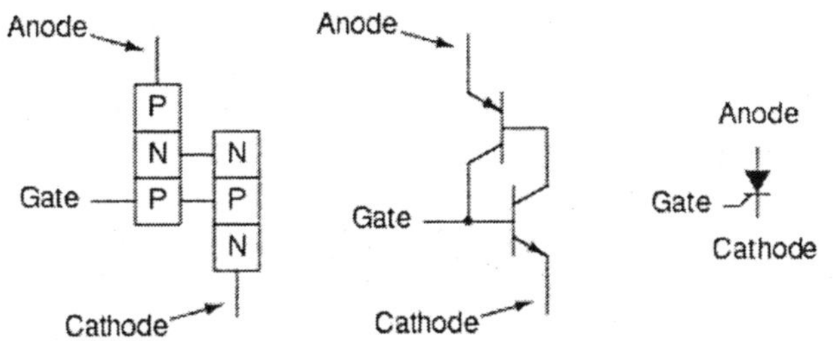

Physical diagram *Equivalent schematic* *Schematic symbol*

If an SCR's gate is left *floating* (disconnected), it behaves exactly as a Shockley diode. It may be latched by breakover voltage or by exceeding the critical rate of voltage rise between anode and cathode, just as with the Shockley diode. Dropout is accomplished by reducing current until one or both internal transistors fall into cutoff mode, also like the Shockley diode. However, because the gate terminal connects directly to the base of the lower transistor, it may be used as an alternative means to latch the SCR.

By applying a small voltage between gate and cathode, the lower transistor will be forced *on* by the resulting base current, which will cause the upper transistor to conduct, which then supplies the lower transistor's base with current so that it no longer needs to be activated by a gate voltage. The necessary gate current to initiate latch-up, of course, will be much lower than the current through the SCR from cathode to anode, so the SCR does achieve a measure of amplification.

This method of securing SCR conduction is called *triggering*, and it is by far the most common way that SCRs are latched in actual practice. In fact, SCRs are usually chosen so that their breakover voltage is far beyond the greatest voltage expected to be experienced from the power source, so that it can be turned on *only* by an intentional voltage pulse applied to the gate.

It should be mentioned that SCRs may *sometimes* be turned off by directly shorting their gate and cathode terminals together, or by "reverse-triggering" the gate with a negative voltage (in reference to the cathode), so that the lower transistor is forced into cutoff. I say this is "sometimes" possible because it involves shunting all of the upper transistor's collector current past the lower transistor's base.

This current may be substantial, making triggered shut-off of an SCR difficult at best. A variation of the SCR, called a *Gate-Turn-Off* thyristor, or *GTO*, makes this task easier. But even with a GTO, the gate current required to turn it off may be as much as 20% of the anode (load) current! The schematic symbol for a GTO is shown in the following illustration:

Gate Turn-Off thyristor (GTO)

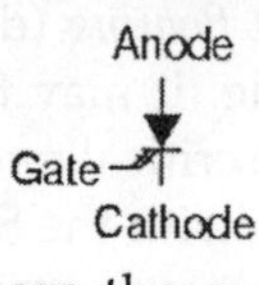

SCRs and GTOs share the same equivalent schematics (two transistors connected in a positive-feedback fashion), the only differences being details of construction designed to grant the NPN transistor a greater β than the PNP.

This allows a smaller gate current (forward or reverse) to exert a greater degree of control over conduction from cathode to anode, with the PNP transistor's latched state being more dependent upon the NPN's than vice versa. The Gate-Turn-Off thyristor is also known by the name of *Gate-Controlled Switch*, or *GCS*.

A rudimentary test of SCR function, or at least terminal identification, may be performed with an ohmmeter. Because the internal connection between gate and cathode is a single

PN junction, a meter should indicate continuity between these terminals with the red test lead on the gate and the black test lead on the cathode like this:

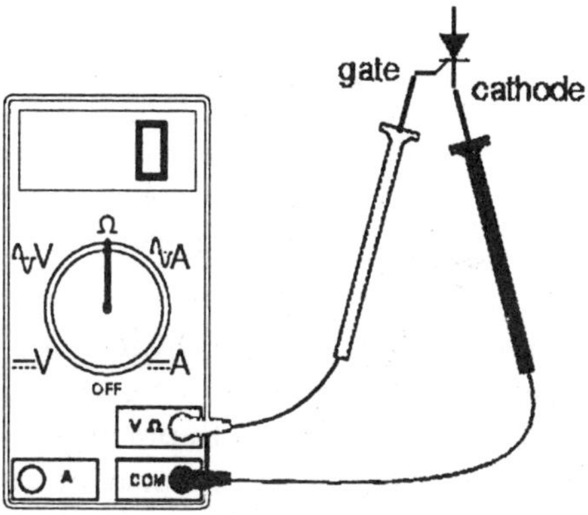

All other continuity measurements performed on an SCR will show "open" ("OL" on some digital multimeter displays). It must be understood that this test is very crude and does *not* constitute a comprehensive assessment of the SCR. It is possible for an SCR to give good ohmmeter indications and still be defective. Ultimately, the only way to test an SCR is to subject it to a load current.

If you are using a multimeter with a "diode check" function, the gate-to-cathode junction voltage indication you get may or may not correspond to what's expected of a silicon PN junction (approximately 0.7 volts). In some cases, you will read a much lower junction voltage: mere hundredths of a volt. This is due to an internal resistor connected between the gate and cathode incorporated within some SCRs.

This resistor is added to make the SCR less susceptible to false triggering by spurious voltage spikes, from circuit "noise" or from static electric discharge. In other words, having a resistor connected across the gate-cathode junction requires that a *strong* triggering signal (substantial current) be applied to latch the SCR.

This feature is often found in larger SCRs, not on small SCRs. Bear in mind that an SCR with an internal resistor connected between gate and cathode will indicate continuity *in both directions* between those two terminals:

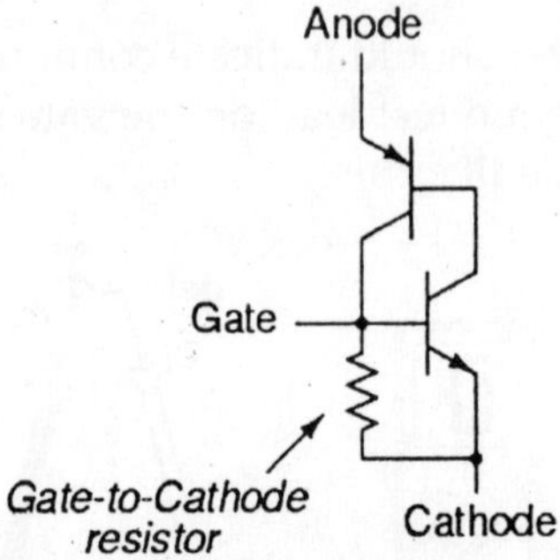

"Normal" SCRs, lacking this internal resistor, are sometimes referred to as *sensitive gate* SCRs due to their ability to be triggered by the slightest positive gate signal.

The test circuit for an SCR is both practical as a diagnostic tool for checking suspected SCRs and also an excellent aid to understanding basic SCR operation. A DC voltage source is used for powering the circuit, and two pushbutton switches are used to latch and unlatch the SCR, respectively:

SCR testing circuit

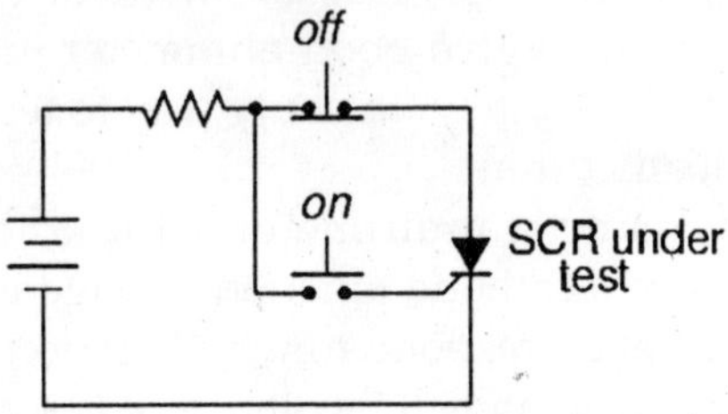

Actuating the normally-open "on" pushbutton switch connects the gate to the anode, allowing current from the negative terminal of the battery, through the cathode-gate PN junction, through the switch, through the load resistor, and back to the battery. This gate current should force the SCR to latch on, allowing current to go directly from cathode to anode without further triggering through the gate. When the "on" pushbutton is released, the load should remain energized.

Pushing the normally-closed "off" pushbutton switch breaks the circuit, forcing current through the SCR to halt, thus forcing it to turn off (low-current dropout).

If the SCR fails to latch, the problem may be with the load and not the SCR. There is a certain minimum amount of load current required to hold the SCR latched in the "on" state. This minimum current level is called the *holding current*. A load with too great a resistance value may not draw enough current to keep an SCR latched when gate current ceases, thus giving the false impression of a bad (unlatchable) SCR in the test circuit. Holding current values for different SCRs should be available from the manufacturers. Typical holding current values range from 1 milliamp to 50 milliamps or more for larger units.

For the test to be fully comprehensive, more than the triggering action needs to be tested. The forward breakover voltage limit of the SCR could be tested by increasing the DC voltage supply (with no pushbuttons actuated) until the SCR latches all on its own.

Beware that a breakover test may require very high voltage: many power SCRs have breakover voltage ratings of 600 volts or more! Also, if a pulse voltage generator is available, the critical rate of voltage rise for the SCR could be tested in the same way: subject it to pulsing supply voltages of different V/time rates with no pushbutton switches actuated and see when it latches.

In this simple form, the SCR test circuit could suffice as a start/stop control circuit for a DC motor, lamp, or other practical load:

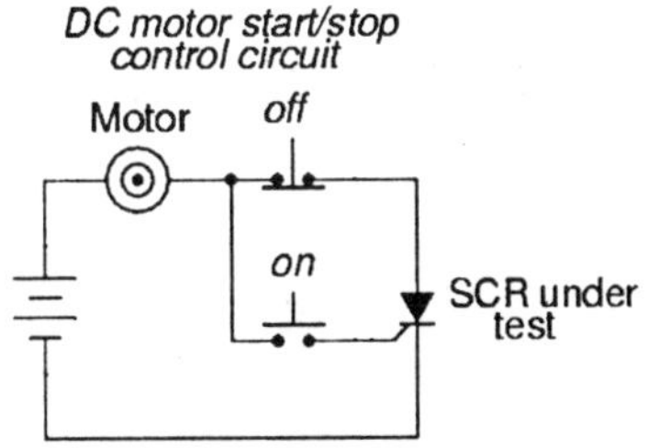

Another practical use for the SCR in a DC circuit is as a *crowbar* device for overvoltage protection. A "crowbar" circuit consists of an SCR placed in parallel with the output of a DC power supply, for the purpose of placing a direct short-circuit on the output of that supply to prevent excessive voltage from reaching the load. Damage to the SCR and power supply is prevented by the judicious placement of a fuse or substantial series resistance ahead of the SCR to limit short-circuit current:

Crowbar as used in an AC-DC power supply

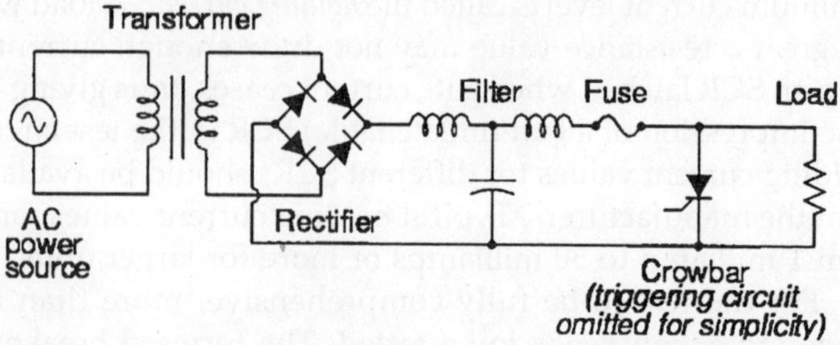

Some device or circuit sensing the output voltage will be connected to the gate of the SCR, so that when an over voltage condition occurs, voltage will be applied between the gate and cathode.

Triggering the SCR and forcing the fuse to blow. The effect will be approximately the same as dropping a solid steel crowbar directly across the output terminals of the power supply, hence the name of the circuit.

Most applications of the SCR are for AC power control, despite the fact that SCRs are inherently DC (unidirectional) devices. If bidirectional circuit current is required, multiple SCRs may be used, with one or more facing each direction to handle current through both half-cycles of the AC wave. The primary reason SCRs are used at all for AC power control applications is the unique response of a thyristor to an alternating current.

As we saw in the case of the thyratron tube (the electron tube version of the SCR) and the DIAC, a hysteretic device triggered on during a portion of an AC half-cycle will latch and remain on throughout the remainder of the half-cycle until the AC current decreases to zero, as it must to begin the next half-cycle.

Just prior to the zero-crossover point of the current waveform, the thyristor will turn off due to insufficient current (this behavior is also known as *natural commutation*) and must be fired again during the next cycle. The result is a circuit current equivalent to a "chopped up" sine wave. For review, here is the graph of a DIAC's response to an AC voltage whose peak exceeds the breakover voltage of the DIAC:

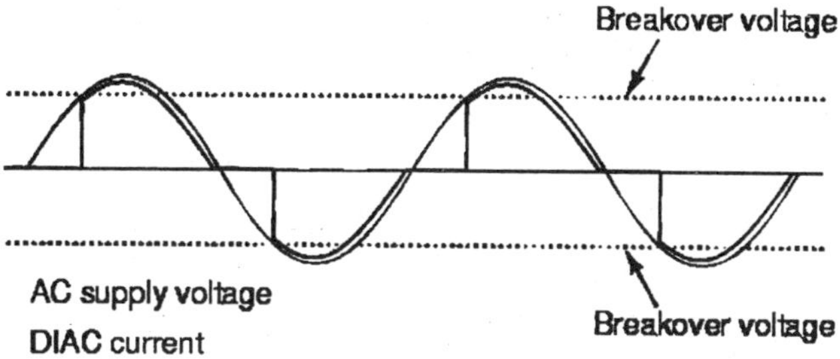

With the DIAC, that breakover voltage limit was a fixed quantity. With the SCR, we have control over exactly when the device becomes latched by triggering the gate at any point in time along the waveform. By connecting a suitable control circuit to the gate of an SCR, we can "chop" the sine wave at any point to allow for time-proportioned power control to a load.

Take the following circuit as an example. Here, an SCR is positioned in a circuit to control power to a load from an AC source:

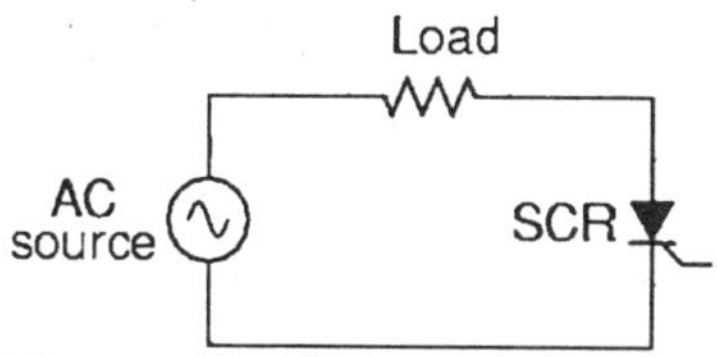

Being a unidirectional (one-way) device, at most we can only deliver half-wave power to the load, in the half-cycle of AC where the supply voltage polarity is positive on the top and negative on the bottom.

However, for demonstrating the basic concept of time-proportional control, this simple circuit is better than one controlling full-wave power (which would require two SCRs). With no triggering to the gate, and the AC source voltage well below the SCR's breakover voltage rating, the SCR will never turn on.

Connecting the SCR gate to the anode through a normal rectifying diode (to prevent reverse current through the gate in the event of the SCR containing a built-in gate-cathode resistor), will allow the SCR to be triggered almost immediately at the beginning of every positive half-cycle:

Gate connected directly to anode through a diode; nearly complete half-wave current through load

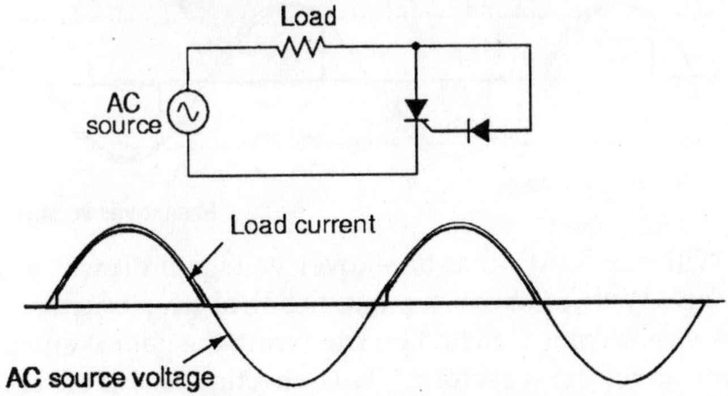

We can delay the triggering of the SCR, however, by inserting some resistance into the gate circuit, thus increasing the amount of voltage drop required before there is enough gate current to trigger the SCR.

In other words, if we make it harder for electrons to flow through the gate by adding a resistance, the AC voltage will have to reach a higher point in its cycle before there will be enough gate current to turn the SCR on. The result looks like this:

Resistance inserted in gate circuit; less than half-wave current through load

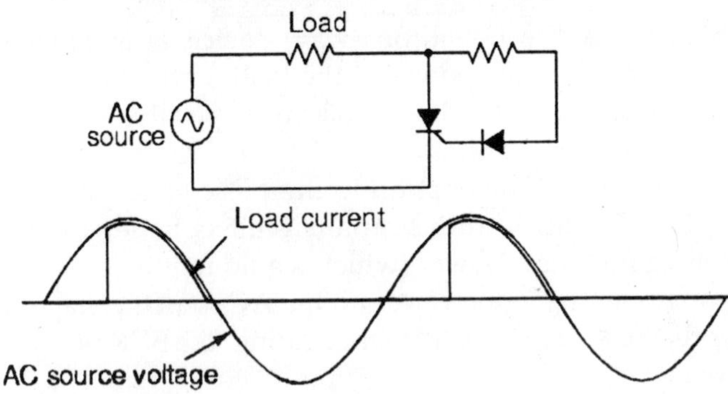

With the half-sine wave chopped up to a greater degree by delayed triggering of the SCR, the load receives less average power (power is delivered for less time throughout a cycle). By making the series gate resistor variable, we can make adjustments to the time-proportioned power:

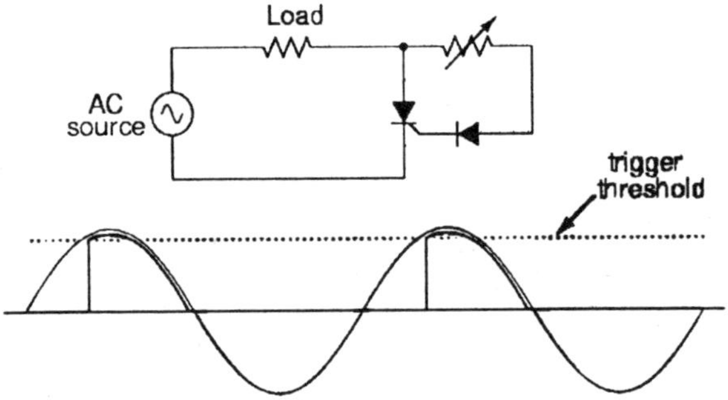

Increasing the resistance raises the threshold level, causing less power to be delivered to the load.

Decreasing the resistance lowers the threshold level, causing more power to be delivered to the load.

Unfortunately, this control scheme has a significant limitation. In using the AC source waveform for our SCR triggering signal, we limit control to the first half of the waveform's half-cycle.

In other words, there is no way for us to wait until *after* the wave's peak to trigger the SCR. This means we can turn down the power only to the point where the SCR turns on at the very peak of the wave:

Circuit at minimum power setting

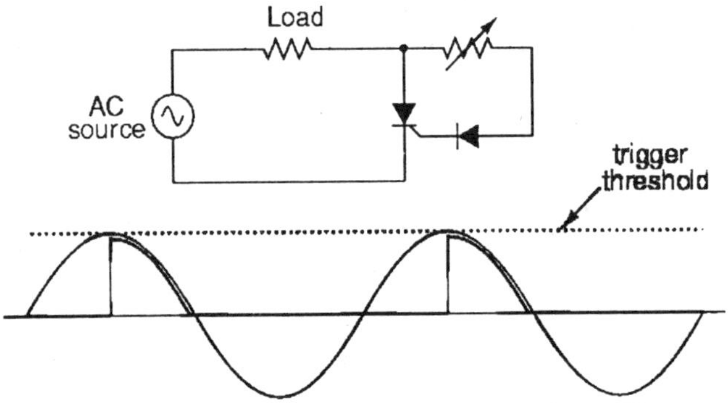

Raising the trigger threshold any more will cause the circuit to not trigger at all, since not even the peak of the AC power voltage will be enough to trigger the SCR. The result will be no power to the load.

An ingenious solution to this control dilemma is found in the addition of a phase-shifting capacitor to the circuit:

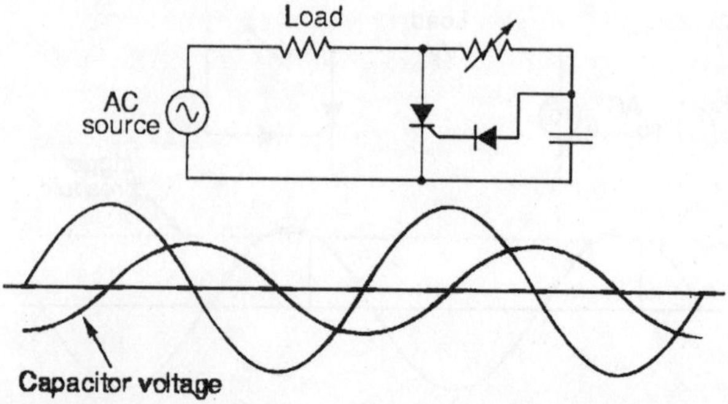

The smaller waveform shown on the graph is voltage across the capacitor. For the sake of illustrating the phase shift, I'm assuming a condition of maximum control resistance where the SCR is not triggering at all and there is no load current, save for what little current goes through the control resistor and capacitor. This capacitor voltage will be phase-shifted anywhere from 0° to 90° lagging behind the power source AC waveform. When this phase-shifted voltage reaches a high enough level, the SCR will trigger.

Assuming there is periodically enough voltage across the capacitor to trigger the SCR, the resulting load current waveform will look something like this:

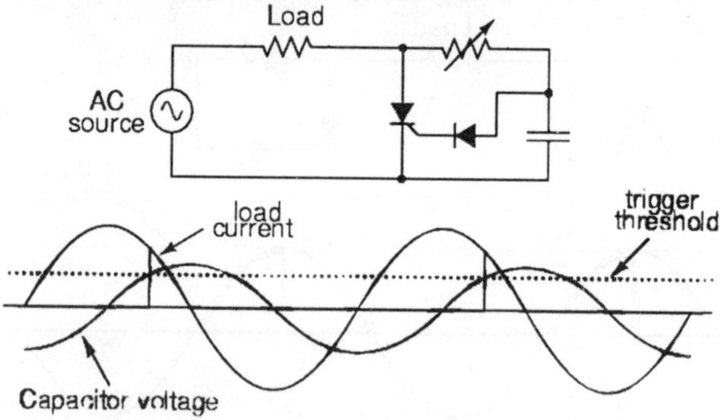

Because the capacitor waveform is still *rising* after the main AC power waveform has reached its peak, it becomes possible to trigger the SCR at a threshold level beyond that peak, thus chopping the load current wave further than it was possible with the simpler circuit. In reality, the capacitor voltage waveform is a bit more complex that what is shown here, its sinusoidal shape distorted every time the SCR latches on. However, what I'm trying to illustrate here is the delayed triggering action gained with the phase-shifting RC network, and so a simplified, undistorted waveform serves the purpose well.

SCRs may also be triggered, or "fired," by more complex circuits. While the circuit previously shown is sufficient for a simple application like a lamp control, large industrial motor controls often rely on more sophisticated triggering methods. Sometimes, pulse transformers are used to couple a triggering circuit to the gate and cathode of an SCR to provide electrical isolation between the triggering and power circuits:

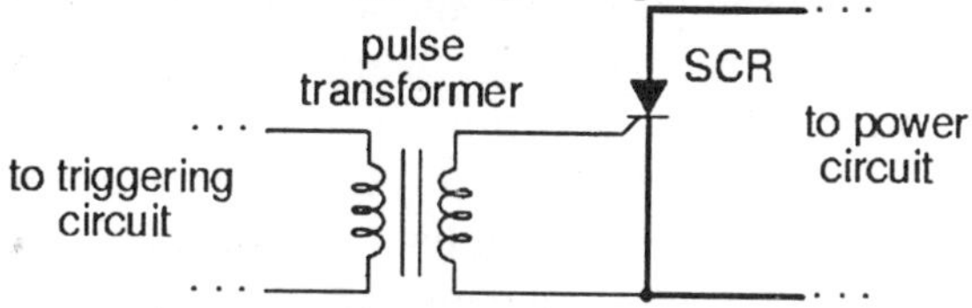

When multiple SCRs are used to control power, their cathodes are often *not* electrically common, making it difficult to connect a single triggering circuit to all SCRs equally. An example of this is the *controlled bridge rectifier* shown here:

Controlled bridge rectifier

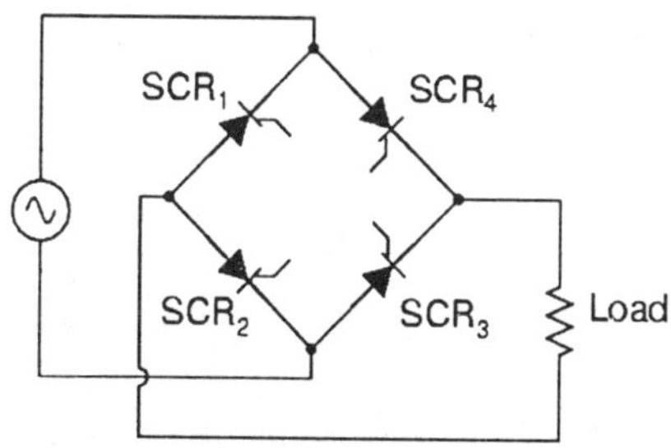

In any bridge rectifier circuit, the rectifying diodes (or in this case, the rectifying SCRs) must conduct in opposite pairs. SCR_1 and SCR_3 must be fired simultaneously, and likewise SCR_2 and SCR_4 must be fired together as a pair. As you will notice, though, these pairs of SCRs do not share the same cathode connections, meaning that it would not work to simply parallel their respective gate connections and connect a single voltage source to trigger both:

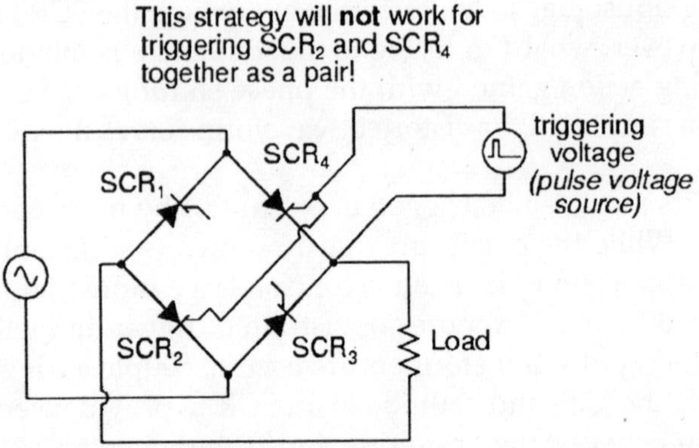

Although the triggering voltage source shown will trigger SCR_4, it will not trigger SCR_2 properly because the two thyristors do not share a common cathode connection to reference that triggering voltage. Pulse transformers connecting the two thyristor gates to a common triggering voltage source *will* work, however:

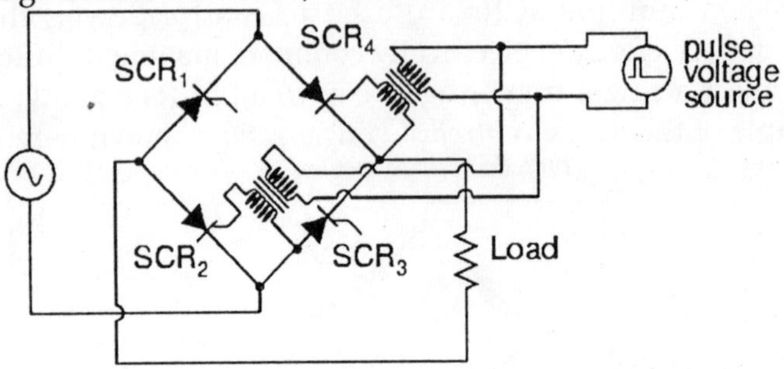

Bear in mind that this circuit only shows the gate connections for two out of the four SCRs. Pulse transformers and triggering sources for SCR_1 and SCR_3, as well as the details of the pulse sources themselves, have been omitted for the sake of simplicity.

Controlled bridge rectifiers are not limited to single-phase designs. In most industrial control systems, AC power is available in three-phase form for maximum efficiency, and solid-state control circuits are built to take advantage of that. A three-phase controlled rectifier circuit built with SCRs, without pulse transformers or triggering circuitry shown, would look like this:

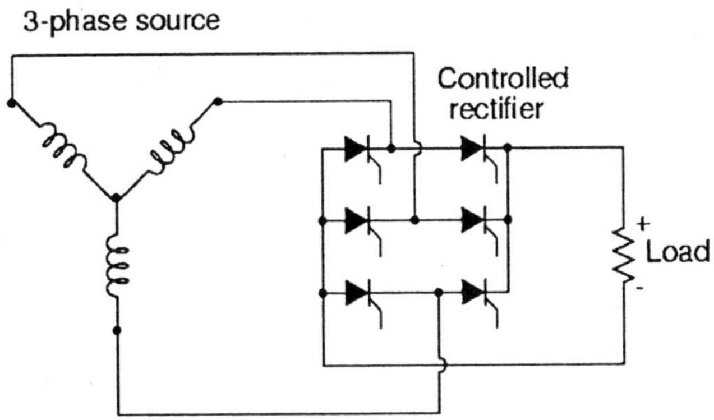

THE TRIAC

SCRs are unidirectional (one-way) current devices, making them useful for controlling DC only. If two SCRs are joined.

In back-to-back parallel fashion just like two Shockley diodes were joined together to form a DIAC, we have a new device known as the *TRIAC*:

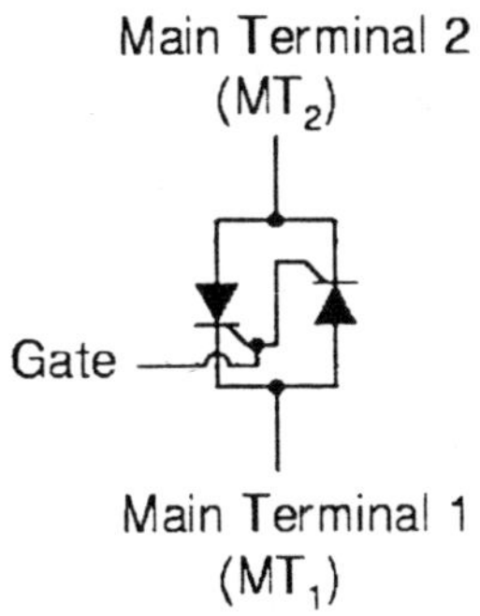

TRIAC equivalent circuit

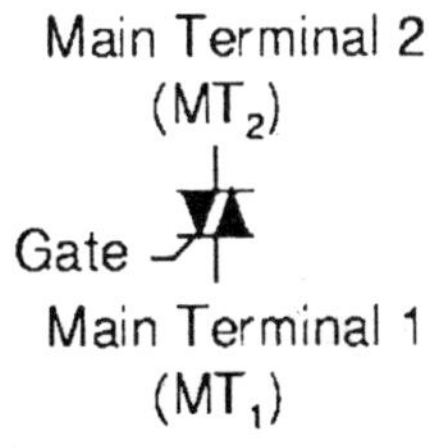

TRIAC schematic symbol

Because individual SCRs are more flexible to use in advanced control systems, they are more commonly seen in circuits like motor drives, while TRIACs are usually seen in simple, low-power applications like household dimmer switches.

A simple lamp dimmer circuit is shown here, complete with the phase-shifting resistor-capacitor network necessary for after-peak firing.

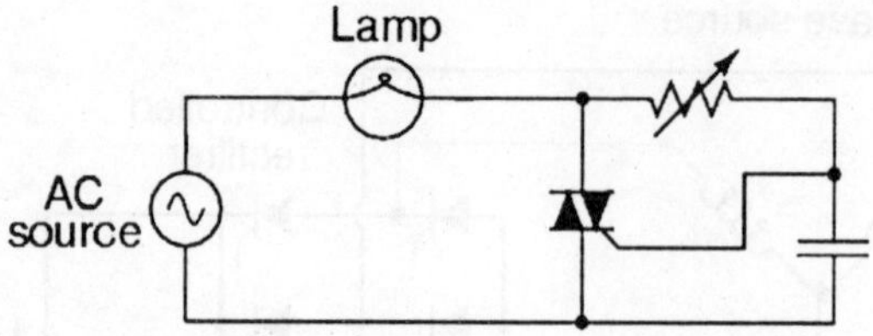

TRIACs are notorious for not firing *symmetrically*. This means they usually won't trigger at the exact same gate voltage level for one polarity as for the other. Generally speaking, this is undesirable, because unsymmetrical firing results in a current waveform with a greater variety of harmonic frequencies. Waveforms that are symmetrical above and below their average centerlines are comprised of only odd-numbered harmonics. Unsymmetrical waveforms, on the other hand, contain even-numbered harmonics (which may or may not be accompanied by odd-numbered harmonics as well).

In the interest of reducing total harmonic content in power systems, the fewer and less diverse the harmonics, the better — one more reason why individual SCRs are favored over TRIACs for complex, high-power control circuits. One way to make the TRIAC's current waveform more symmetrical is to use a device external to the TRIAC to time the triggering pulse. A DIAC placed in series with the gate does a fair job of this:

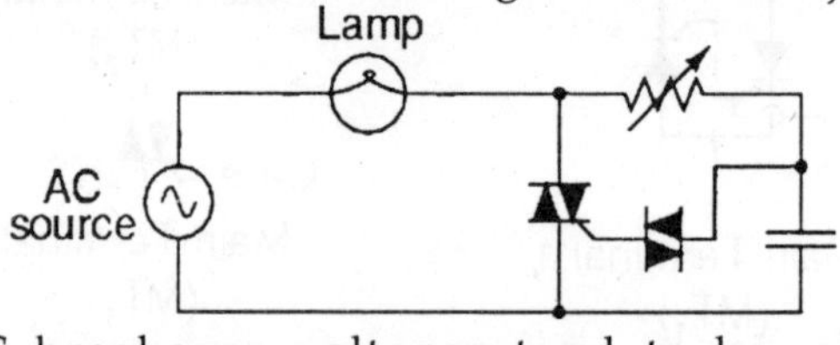

DIAC breakover voltages tend to be much more symmetrical (the same in one polarity as the other) than TRIAC triggering voltage thresholds. Since the DIAC prevents any gate current until the triggering voltage has reached a certain, repeatable level in either direction, the firing point of the TRIAC from one half-cycle to the next tends to be more consistent, and the waveform more symmetrical above and below its centerline.

Practically all the characteristics and ratings of SCRs apply equally to TRIACs, except that TRIACs of course are bidirectional (can handle current in both directions). Not much more needs to be said about this device except for an important caveat concerning its terminal designations.

From the equivalent circuit diagram shown earlier, one might think that main terminals 1 and 2 were interchangeable. They are not! Although it is helpful to imagine the TRIAC as being composed of two SCRs joined together, it in fact is constructed from a single piece of semiconducting material, appropriately doped and layered. The actual operating characteristics may differ slightly from that of the equivalent model.

This is made most evident by contrasting two simple circuit designs, one that works and one that doesn't. The following two circuits are a variation of the lamp dimmer circuit shown earlier, the phase-shifting capacitor and DIAC removed for simplicity's sake. Although the resulting circuit lacks the fine control ability of the more complex version (with capacitor and DIAC), it *does* function:

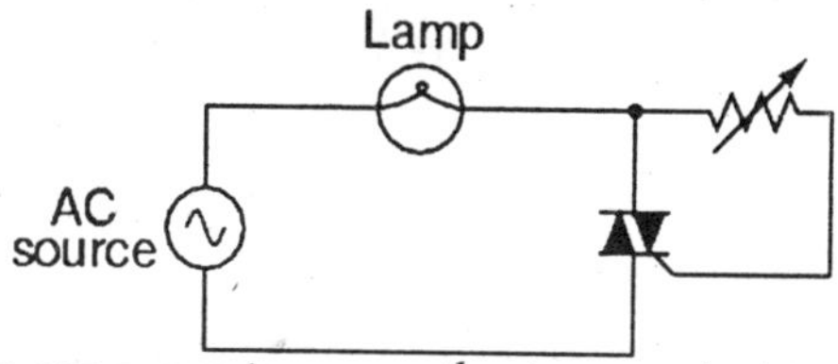

Suppose we were to swap the two main terminals of the TRIAC around. According to the equivalent circuit diagram shown earlier in this section, the swap should make no difference. The circuit ought to work:

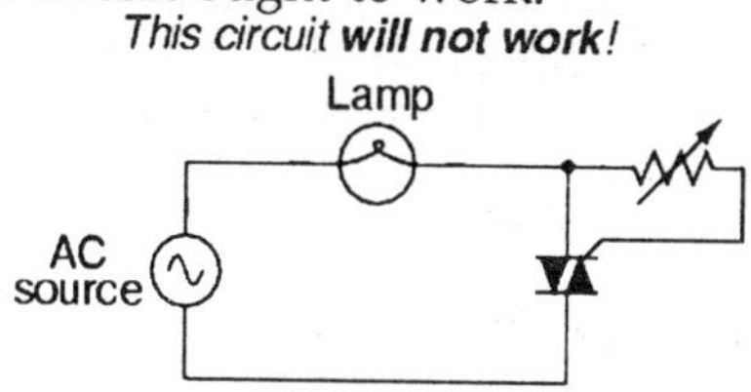

However, if this circuit is built, it will be found that it does not work! The load will receive no power, the TRIAC refusing to fire at all, no matter how low or high a resistance value the control resistor is set to. The key to successfully triggering a TRIAC is to make sure the gate receives its triggering current from the *main terminal* 2 side of the circuit (the main terminal on the opposite side of the TRIAC symbol from the gate terminal).

OPTOTHYRISTORS

Like bipolar transistors, SCRs and TRIACs are also manufactured as light-sensitive devices, the action of impinging light replacing the function of triggering voltage.

Optically-controlled SCRs are often known by the acronym *LASCR*, or Light Activated SCR. Its symbol, not surprisingly, looks like this:

Light Activated SCR

LASCR

Optically-controlled TRIACs don't receive the honor of having their own acronym, but instead are humbly known as opto-TRIACs. Their schematic symbol looks like this:

Opto-TRIAC

Optothyristors (a general term for either the LASCR or the opto-TRIAC) are commonly found inside sealed "optoisolator" modules.

THE SILICON-CONTROLLED SWITCH (SCS)

If we take the equivalent circuit for an SCR and add another external terminal, connected to the base of the top transistor and the collector of the bottom transistor, we have a device known as a *silicon-controlled-switch*, or *SCS*:

The Silicon-Controlled Switch (SCS)

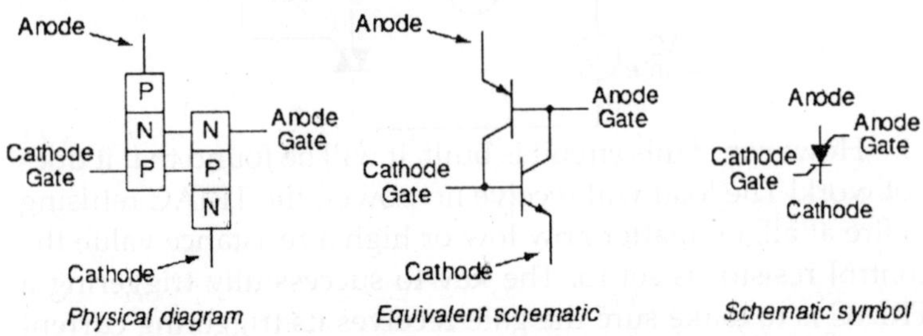

This extra terminal allows more control to be exerted over the device, particularly in the mode of *forced commutation,* where an external signal forces it to turn off while the main current through the device has not yet fallen below the holding current value. Consider the following circuit:

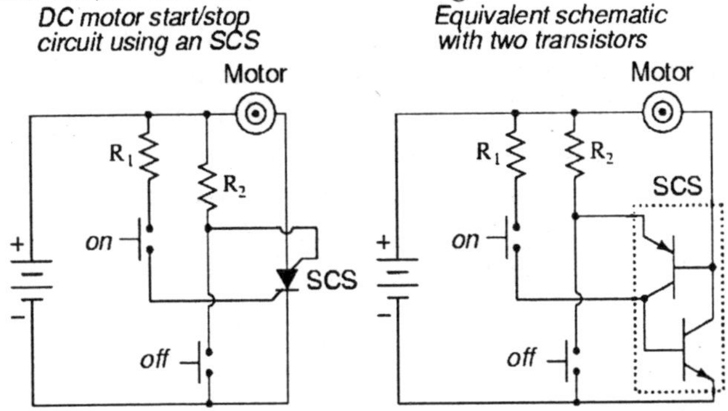

When the "on" pushbutton switch is actuated, there is a voltage applied between the cathode gate and the cathode, forward-biasing the lower transistor's base-emitter junction, and turning it on. The top transistor of the SCS is ready to conduct, having been supplied with a current path from its emitter terminal (the SCS's anode terminal) through resistor R_2 to the positive side of the power supply. As in the case of the SCR, both transistors turn on and maintain each other in the "on" mode. When the lower transistor turns on, it conducts the motor's load current, and the motor starts and runs.

The motor may be stopped by interrupting the power supply, as with an SCR, and this is called *natural commutation.* However, the SCS provides us with another means of turning off: *forced commutation* by shorting the anode terminal to the cathode. If this is done (by actuating the "off" pushbutton switch), the upper transistor within the SCS will lose its emitter current, thus halting current through the base of the lower transistor.

When the lower transistor turns off, it breaks the circuit for base current through the top transistor (securing its "off" state), and the motor (making it stop). The SCS will remain in the off condition until such time that the "on" pushbutton switch is re-actuated.

FIELD-EFFECT-CONTROLLED THYRISTORS

Two relatively recent technologies designed to reduce the "driving" (gate trigger current) requirements of classic thyristor devices are the *MOS-gated thyristor* and the *MOS Controlled Thyristor*, or *MCT*.

The MOS-gated thyristor uses a MOSFET to initiate conduction through the upper (PNP) transistor of a normal thyristor structure, thus triggering the device. Since a MOSFET requires negligible current to "drive" (cause it to saturate), this makes the thyristor as a whole very easy to trigger:

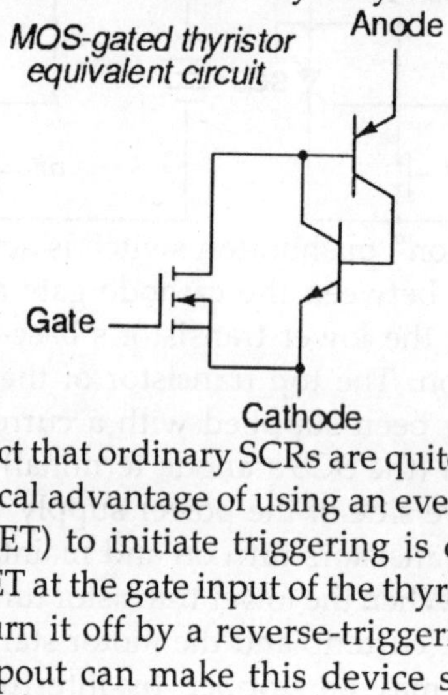

Given the fact that ordinary SCRs are quite easy to "drive" as it is, the practical advantage of using an even more sensitive device (a MOSFET) to initiate triggering is debatable. Also, placing a MOSFET at the gate input of the thyristor now makes it *impossible* to turn it off by a reverse-triggering signal. Only low-current dropout can make this device stop conducting after it has been latched.

A device of arguably greater value would be a fully-controllable thyristor, whereby a small gate signal could both trigger the thyristor and force it to turn off. Such a device does exist, and it is called the *MOS Controlled Thyristor*, or *MCT*. It uses a pair of MOSFETs connected to a common gate terminal, one to trigger the thyristor and the other to "untrigger" it:

A positive gate voltage (with respect to the cathode) turns on the upper (N-channel) MOSFET, allowing base current through the upper (PNP) transistor, which latches the transistor pair in an "on" state. Once both transistors are fully latched, there will be little voltage dropped between anode and cathode, and the thyristor will remain latched so long as the controlled current exceeds the minimum (holding) current value.

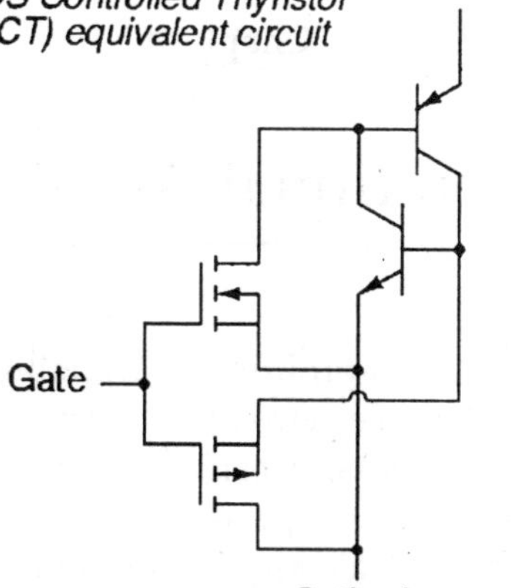

However, if a negative gate voltage is applied (with respect to the anode, which is at nearly the same voltage as the cathode in the latched state), the lower MOSFET will turn on and "short" between the lower (NPN) transistor's base and emitter terminals, thus forcing it into cutoff. Once the NPN transistor cuts off, the PNP transistor will drop out of conduction, and the whole thyristor turns off. Gate voltage has full control over conduction through the MCT: to turn it on and to turn it off.

This device is still a thyristor, though. If there is zero voltage applied between gate and cathode, neither MOSFET will turn on. Consequently, the bipolar transistor pair will remain in whatever state it was last in (hysteresis). So, a brief positive pulse to the gate turns the MCT on, a brief negative pulse forces it off, and no applied gate voltage lets it remain in whatever state it is already in. In essence, the MCT is a latching version of the IGBT (Insulated Gate Bipolar Transistor).

Chapter 6

Operational Amplifiers

The operational amplifier is arguably the most useful single device in analog electronic circuitry. With only a handful of external components, it can be made to perform a wide variety of analog signal processing tasks. It is also quite affordable, most general-purpose amplifiers selling for under a dollar apiece.

Modern designs have been engineered with durability in mind as well:

Several "op-amps" are manufactured that can sustain direct short-circuits on their outputs without damage.

One key to the usefulness of these little circuits is in the engineering principle of feedback, particularly *negative* feedback, which constitutes the foundation of almost all automatic control processes.

The principles presented here in operational amplifier circuits, therefore, extend well beyond the immediate scope of electronics.

It is well worth the electronics student's time to learn these principles and learn them well.

SINGLE-ENDED AND DIFFERENTIAL AMPLIFIERS

For ease of drawing complex circuit diagrams, electronic amplifiers are often symbolized by a simple triangle shape, where the internal components are not individually represented.

This symbology is very handy for cases where an amplifier's construction is irrelevant to the greater function of the overall circuit, and it is worthy of familiarization:

General amplifier circuit symbol

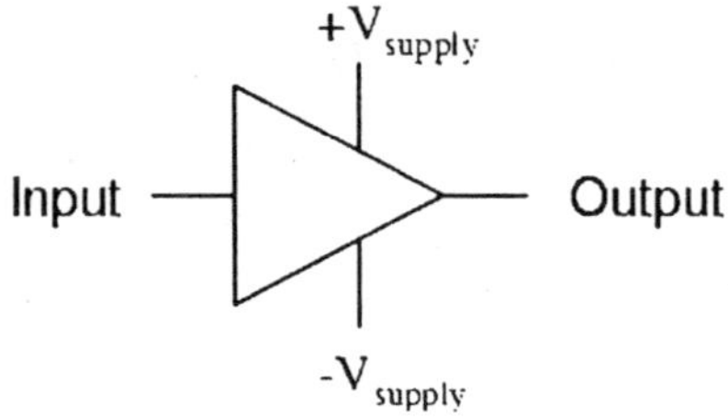

The +V and -V connections denote the positive and negative sides of the DC power supply, respectively. The input and output voltage connections are shown as single conductors, because it is assumed that all signal voltages are referenced to a common connection in the circuit called *ground*. Often (but not always!), one pole of the DC power supply, either positive or negative, is that ground reference point. A practical amplifier circuit (showing the input voltage source, load resistance, and power supply) might look like this:

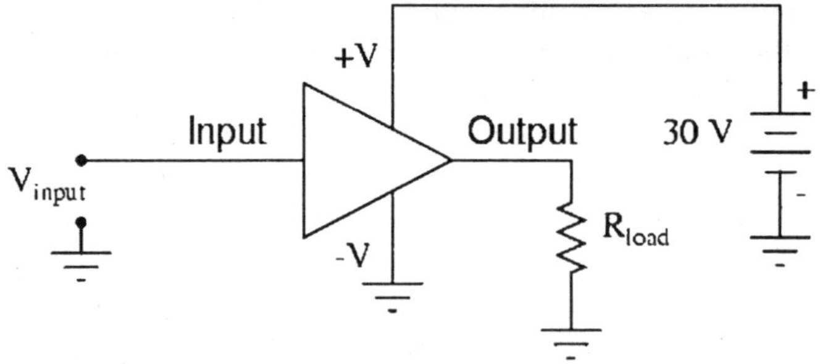

Without having to analyze the actual transistor design of the amplifier, you can readily discern the whole circuit's function: to take an input signal (V_{in}), amplify it, and drive a load resistance (R_{load}).

To complete the above schematic, it would be good to specify the gains of that amplifier (A_V, A_I, A_P) and the Q (bias) point for any needed mathematical analysis.

If it is necessary for an amplifier to be able to output true AC voltage (reversing polarity) to the load, a *split* DC power supply may be used, whereby the ground point is electrically "centered" between +V and -V. Sometimes the split power supply configuration is referred to as a *dual* power supply.

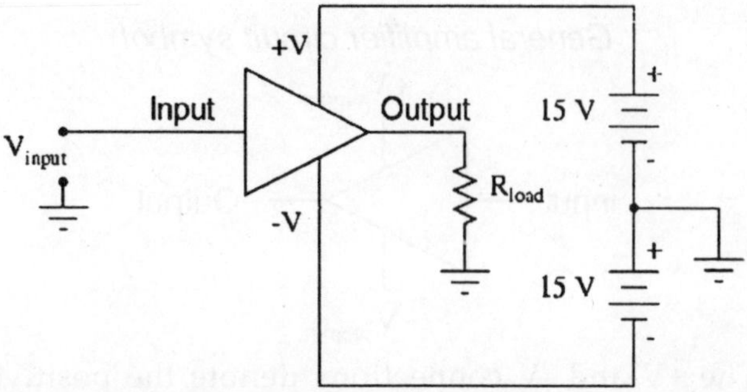

The amplifier is still being supplied with 30 volts overall, but with the split voltage DC power supply, the output voltage across the load resistor can now swing from a theoretical maximum of +15 volts to -15 volts, instead of +30 volts to 0 volts. This is an easy way to get true alternating current (AC) output from an amplifier without resorting to capacitive or inductive (transformer) coupling on the output. The peak-to-peak amplitude of this amplifier's output between cutoff and saturation remains unchanged.

By signifying a transistor amplifier within a larger circuit with a triangle symbol, we ease the task of studying and analyzing more complex amplifiers and circuits. One of these more complex amplifier types that we'll be studying is called the *differential amplifier*. Unlike normal amplifiers, which amplify a single input signal (often called *single-ended* amplifiers), differential amplifiers amplify the voltage difference between two input signals. Using the simplified triangle amplifier symbol, a differential amplifier looks like this:

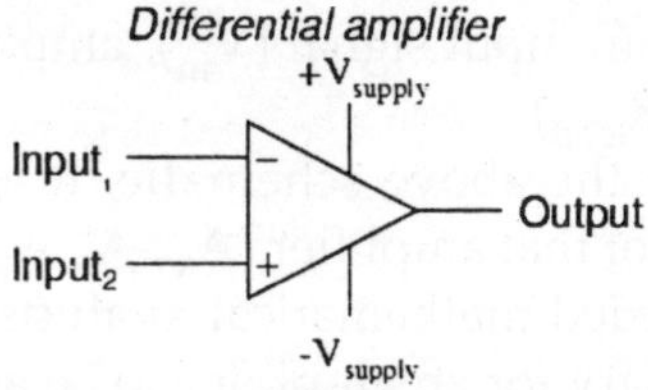

The two input leads can be seen on the left-hand side of the triangular amplifier symbol, the output lead on the right-hand side, and the +V and -V power supply leads on top and bottom. As with the other example, all voltages are referenced to the circuit's ground point. Notice that one input lead is marked with a (-) and the other is marked with a (+). Because a differential amplifier amplifies the difference in voltage between the two inputs, each input influences the output voltage in opposite ways. Consider the following table of input/output voltages for a differential amplifier with a voltage gain of 4:

(–) Input$_1$	0	0	0	0	1	2.5	7	3	–3	2
(+) Input$_2$	0	1	2.5	7	0	0	0	3	3	–7
output	0	4	10	28	–4	–10	–28	0	24	–20

Voltage Output Equation:

$$V_{out} = A_V(\text{Input}_2 - \text{Input}_1)$$

$$V_{out} = A_V(\text{Input}_{(+)} - \text{Input}_{(-)})$$

An increasingly positive voltage on the (+) input tends to drive the output voltage more positive, and an increasingly positive voltage on the (-) input tends to drive the output voltage more negative. Likewise, an increasingly negative voltage on the (+) input tends to drive the output negative as well, and an increasingly negative voltage on the (-) input does just the opposite. Because of this relationship between inputs and polarities, the (-) input is commonly referred to as the *inverting* input and the (+) as the *noninverting* input.

It may be helpful to think of a differential amplifier as a variable voltage source controlled by a sensitive voltmeter, as such:

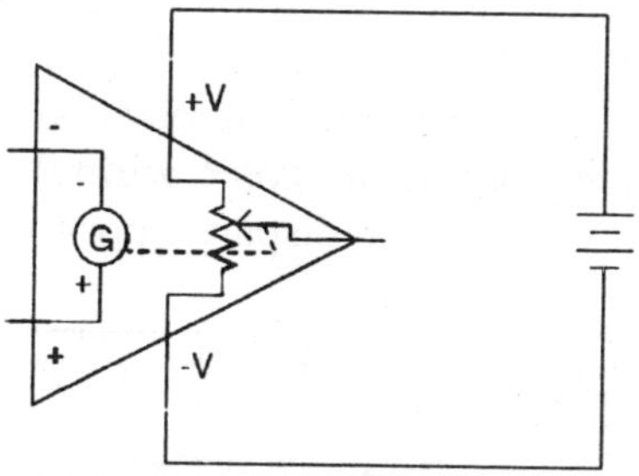

Bear in mind that the above illustration is only a *model* to aid in understanding the behavior of a differential amplifier. It is not a realistic schematic of its actual design. The "G" symbol represents a galvanometer, a sensitive voltmeter movement. The potentiometer connected between +V and -V provides a variable voltage at the output pin (with reference to one side of the DC power supply), that variable voltage set by the reading of the galvanometer. It must be understood that any load powered by the output of a differential amplifier gets its current from the DC power source (battery), *not* the input signal. The input signal (to the galvanometer) merely *controls* the output.

This concept may at first be confusing to students new to amplifiers. With all these polarities and polarity markings (- and +) around, its easy to get confused and not know what the output of a differential amplifier will be. To address this potential confusion, here's a simple rule to remember:

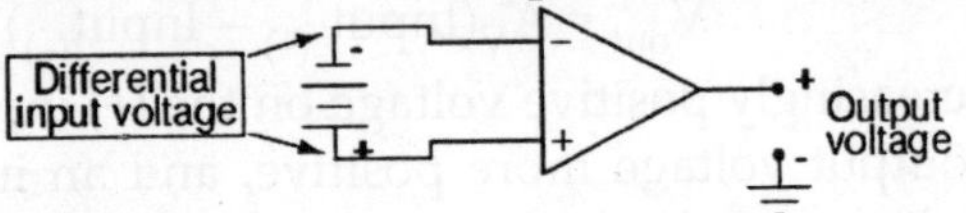

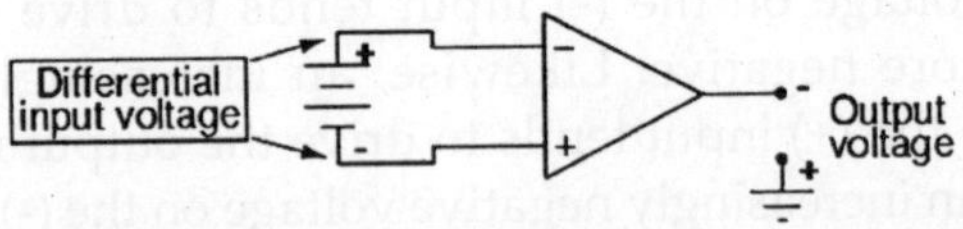

When the polarity of the *differential* voltage matches the markings for inverting and noninverting inputs, the output will be positive. When the polarity of the differential voltage clashes with the input markings, the output will be negative. This bears some similarity to the mathematical sign displayed by digital voltmeters based on input voltage polarity. The red test lead of the voltmeter (often called the "positive" lead because of the color red's popular association with the positive side of a power supply in electronic wiring) is more positive than the black, the meter will display a positive voltage figure, and vice versa:

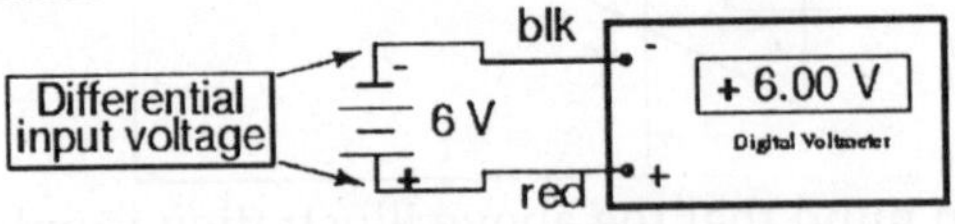

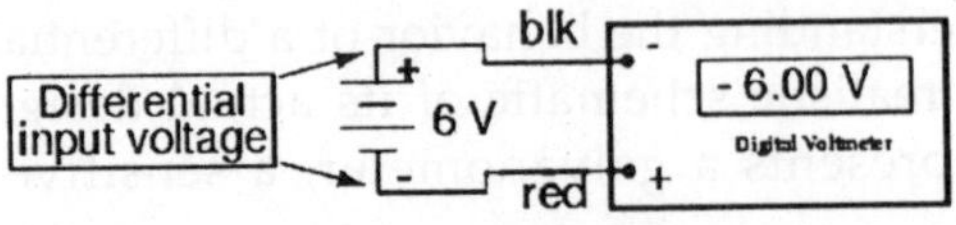

Just as a voltmeter will only display the voltage *between* its two test leads, an ideal differential amplifier only amplifies the potential difference between its two input connections, not the voltage between any one of those connections and ground. The output polarity of a differential amplifier, just like the signed indication of a digital voltmeter, depends on the relative polarities of the differential voltage between the two input connections.

If the input voltages to this amplifier represented mathematical quantities (as is the case within analog computer circuitry), or physical process measurements (as is the case within analog electronic instrumentation circuitry), you can see how a device such as a differential amplifier could be very useful. We could use it to compare two quantities to see which is greater (by the polarity of the output voltage), or perhaps we could compare the difference between two quantities (such as the level of liquid in two tanks) and flag an alarm (based on the absolute value of the amplifier output) if the difference became too great.

In basic automatic control circuitry, the quantity being controlled (called the *process variable*) is compared with a target value (called the *set point*), and decisions are made as to how to act based on the discrepancy between these two values. The first step in electronically controlling such a scheme is to amplify the difference between the process variable and the set point with a differential amplifier. In simple controller designs, the output of this differential amplifier can be directly utilized to drive the final control element (such as a valve) and keep the process reasonably close to set point.

THE "OPERATIONAL" AMPLIFIER

Long before the advent of digital electronic technology, computers were built to electronically perform calculations by employing voltages and currents to represent numerical quantities. This was especially useful for the simulation of physical processes. A variable voltage, for instance, might represent velocity or force in a physical system. Through the use of resistive voltage dividers and voltage amplifiers, the mathematical operations of division and multiplication could be easily performed on these signals.

The reactive properties of capacitors and inductors lend themselves well to the simulation of variables related by calculus functions. Remember how the current through a capacitor was a function of the voltage's rate of change, and how that rate of change was designated in calculus as the *derivative*? Well, if voltage across a capacitor were made to represent the velocity of an object, the current through the capacitor would represent the force required to accelerate or decelerate that object, the capacitor's capacitance representing the object's mass:

$$i_C = C\frac{dv}{dt} \qquad F = m\frac{dv}{dt}$$

Where,	Where,
i_C = Instantaneous current through capacitor	F = Force applied to ojbect
C = Capacitance in farads	m = Mass of object
$\frac{dv}{dt}$ = Rate of change of voltage over time	$\frac{dv}{dt}$ = Rate of change of velocity over time.

This analog electronic computation of the calculus derivative function is technically known as *differentiation,* and it is a natural function of a capacitor's current in relation to the voltage applied across it. Note that this circuit requires no "programming" to perform this relatively advanced mathematical function as a digital computer would.

Electronic circuits are very easy and inexpensive to create compared to complex physical systems, so this kind of analog electronic simulation was widely used in the research and development of mechanical systems. For realistic simulation, though, amplifier circuits of high accuracy and easy configurability were needed in these early computers.

It was found in the course of analog computer design that differential amplifiers with extremely high voltage gains met these requirements of accuracy and configurability better than single-ended amplifiers with custom-designed gains. Using simple components connected to the inputs and output of the high-gain differential amplifier, virtually any gain and any function could be obtained from the circuit, overall, without adjusting or modifying the internal circuitry of the amplifier itself. These high-gain differential amplifiers came to be known as *operational amplifiers*, or *op-amps*, because of their application in analog computers' mathematical *operations*.

Modern op-amps, like the popular model 741, are high-performance, inexpensive integrated circuits. Their input impedances are quite high, the inputs drawing currents in the range of half a microamp (maximum) for the 741, and far less for op-amps utilizing field-effect input transistors.

Output impedance is typically quite low, about 75 Ω for the model 741, and many models have built-in output short circuit protection, meaning that their outputs can be directly shorted to ground without causing harm to the internal circuitry. With direct coupling between op-amps' internal transistor stages, they can amplify DC signals just as well as AC (up to certain maximum voltage-risetime limits). It would cost far more in money and time to design a comparable discrete-transistor amplifier circuit to match that kind of performance, unless high power capability was required. For these reasons, op-amps have all but obsoleted discrete-transistor signal amplifiers in many applications.

Typical 8-pin "DIP" op-amp integrated circuit

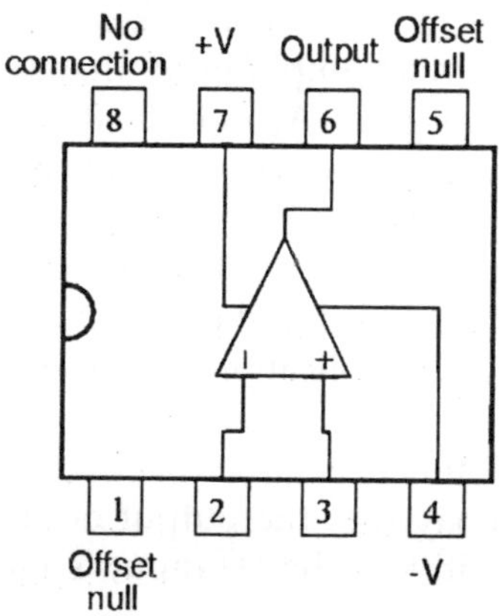

The following diagram shows the pin connections for single op-amps (741 included) when housed in an 8-pin DIP (Dual Inline Package) integrated circuit:

Some models of op-amp come two to a package, including the popular models TL082 and 1458. These are called "dual" units, and are typically housed in an 8-pin DIP package as well, with the following pin connections:

Dual op-amp in 8-pin DIP

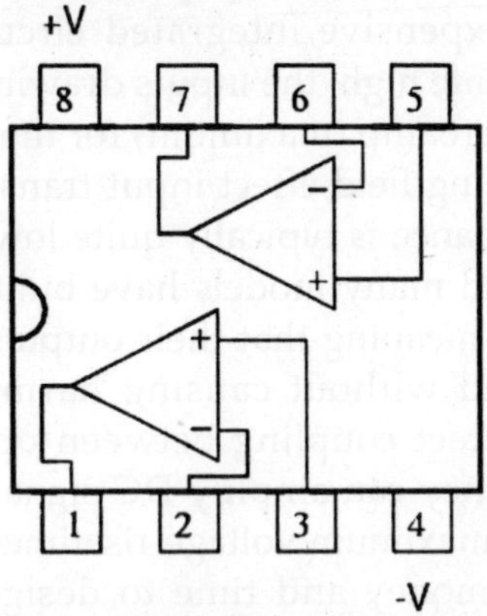

Operational amplifiers are also available four to a package, usually in 14-pin DIP arrangements. Unfortunately, pin assignments aren't as standard for these "quad" op-amps as they are for the "dual" or single units. Consult the manufacturer datasheet(s) for details.

Practical operational amplifier voltage gains are in the range of 200,000 or more, which makes them almost useless as an analog differential amplifier by themselves. For an op-amp with a voltage gain (A_V) of 200,000 and a maximum output voltage swing of +15V/-15V, all it would take is a differential input voltage of 75 μV (microvolts) to drive it to saturation or cutoff! Before we take a look at how external components are used to bring the gain down to a reasonable level, let's investigate applications for the "bare" op-amp by itself.

One application is called the *comparator*. For all practical purposes, we can say that the output of an op-amp will be saturated fully positive if the (+) input is more positive than the (–) input, and saturated fully negative if the (+) input is less positive than the (–) input. In other words, an op-amp's extremely high voltage gain makes it useful as a device to compare two voltages and change output voltage states when one input exceeds the other in magnitude.

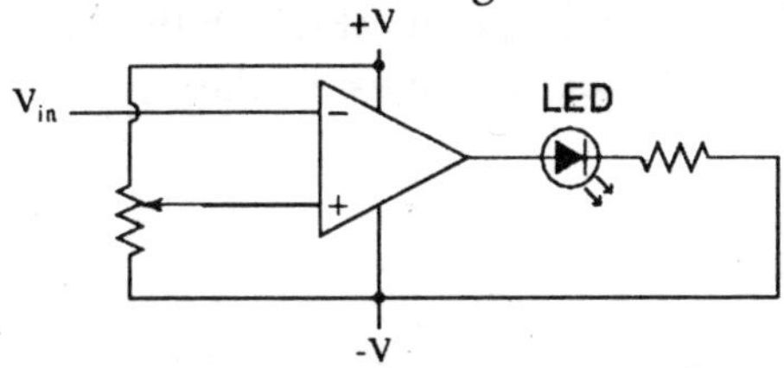

In the above circuit, we have an op-amp connected as a comparator, comparing the input voltage with a reference voltage set by the potentiometer (R_1). If V_{in} drops below the voltage set by R_1, the op-amp's output will saturate to +V, thereby lighting up the LED. Otherwise, if V_{in} is above the reference voltage, the LED will remain off. If V_{in} is a voltage signal produced by a measuring instrument, this comparator circuit could function as a "low" alarm, with the trip-point set by R_1. Instead of an LED, the op-amp output could drive a relay, a transistor, an SCR, or any other device capable of switching power to a load such as a solenoid valve, to take action in the event of a low alarm.

Another application for the comparator circuit shown is a square-wave converter.

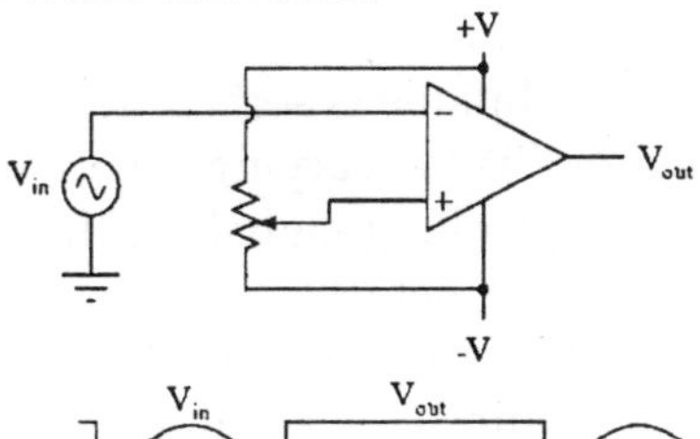

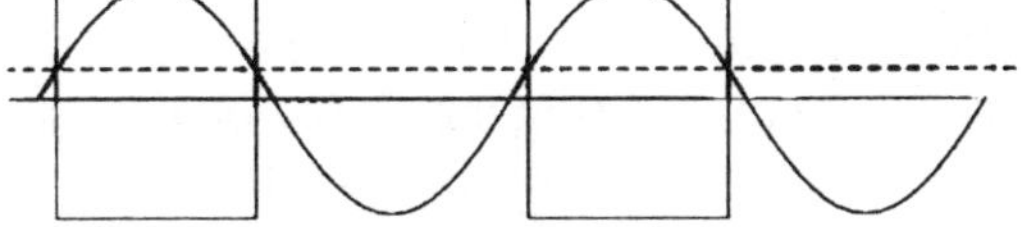

Suppose that the input voltage applied to the inverting (–) input was an AC sine wave rather than a stable DC voltage. In that case, the output voltage would transition between opposing states of saturation whenever the input voltage was equal to the reference voltage produced by the potentiometer. The result would be a square wave:

Adjustments to the potentiometer setting would change the reference voltage applied to the noninverting (+) input, which would change the points at which the sine wave would cross, changing the on/off times, or *duty cycle* of the square wave:

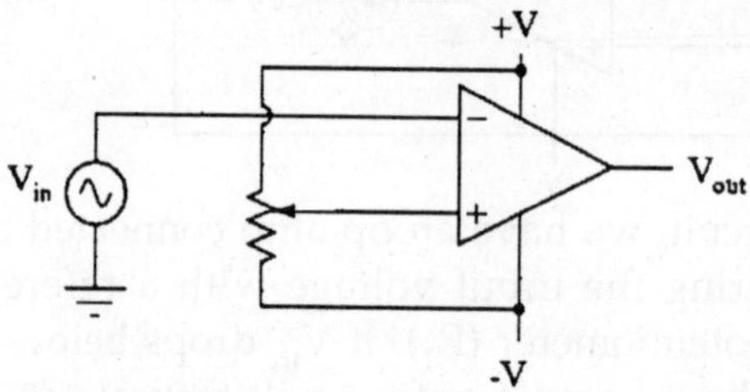

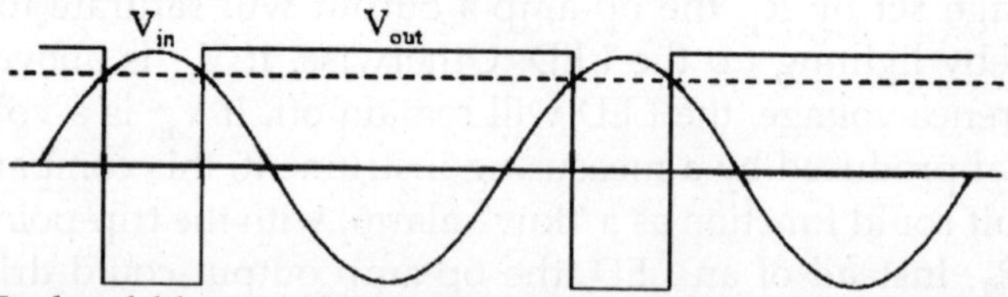

It should be evident that the AC input voltage would not have to be a sine wave in particular for this circuit to perform the same function. The input voltage could be a triangle wave, sawtooth wave, or any other sort of wave that ramped smoothly from positive to negative to positive again. This sort of comparator circuit is very useful for creating square waves of varying duty cycle. This technique is sometimes referred to as *pulse-width modulation*, or PWM (varying, or *modulating* a waveform according to a controlling signal, in this case the signal produced by the potentiometer).

Another comparator application is that of the bargraph driver. If we had several op-amps connected as comparators, each with its own reference voltage connected to the inverting input, but each one monitoring the same voltage signal on their noninverting inputs, we could build a bargraph-style meter such as what is commonly seen on the face of stereo tuners and graphic equalizers. As the signal voltage (representing radio signal strength or audio sound level) increased, each comparator would "turn on" in sequence and send power to its respective LED. With each comparator switching "on" at a different level of audio sound, the number of LED's illuminated would indicate how strong the signal was.

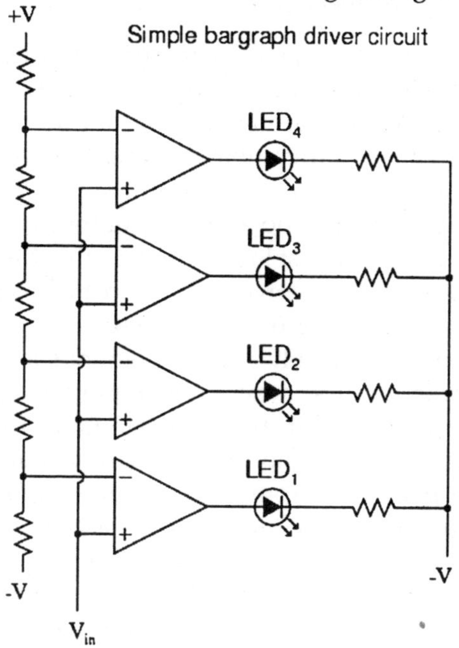

In the circuit shown above, LED_1 would be the first to light up as the input voltage increased in a positive direction. As the input voltage continued to increase, the other LED's would illuminate in succession, until all were lit.

This very same technology is used in some analog-to-digital signal converters, namely the *flash converter*, to translate an analog signal quantity into a series of on/off voltages representing a digital number.

NEGATIVE FEEDBACK

If we connect the output of an op-amp to its inverting input and apply a voltage signal to the noninverting input, we find that the output voltage of the op-amp closely follows that input voltage (I've neglected to draw in the power supply, +V/-V wires, and ground symbol for simplicity):

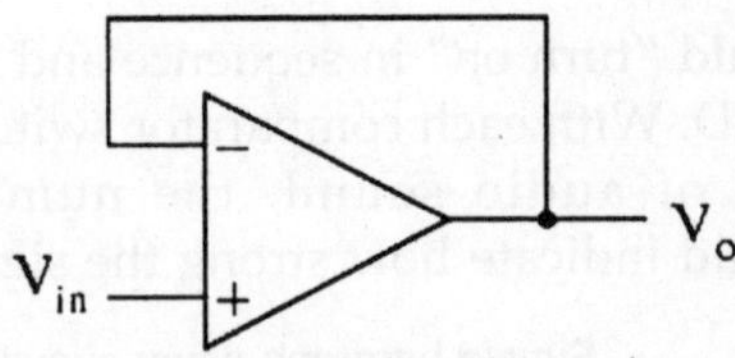

As V_{in} increases, V_{out} will increase in accordance with the differential gain. However, as V_{out} increases, that output voltage is fed back to the inverting input, thereby acting to decrease the voltage differential between inputs, which acts to bring the output down. What will happen for any given voltage input is that the op-amp will output a voltage very nearly equal to V_{in}, but just low enough so that there's enough voltage difference left between V_{in} and the (-) input to be amplified to generate the output voltage.

The circuit will quickly reach a point of stability (known as *equilibrium* in physics), where the output voltage is just the right amount to maintain the right amount of differential, which in turn produces the right amount of output voltage. Taking the op-amp's output voltage and coupling it to the inverting input is a technique known as *negative feedback*, and it is the key to having a self-stabilizing system (this is true not only of op-amps, but of any dynamic system in general). This stability gives the op-amp the capacity to work in its linear (active) mode, as opposed to merely being saturated fully "on" or "off" as it was when used as a comparator, with no feedback at all.

Because the op-amp's gain is so high, the voltage on the inverting input can be maintained almost equal to V_{in}. Let's say that our op-amp has a differential voltage gain of 200,000. If V_{in} equals 6 volts, the output voltage will be 5.999970000149999 volts. This creates just enough differential voltage (6 volts - 5.999970000149999 volts = 29.99985 μV) to cause 5.999970000149999 volts to be manifested at the output terminal, and the system holds there in balance. As you can see, 29.99985 μV is not a lot of differential, so for practical calculations, we can assume that the differential voltage between the two input wires is held by negative feedback exactly at 0 volts.

The effects of negative feedback

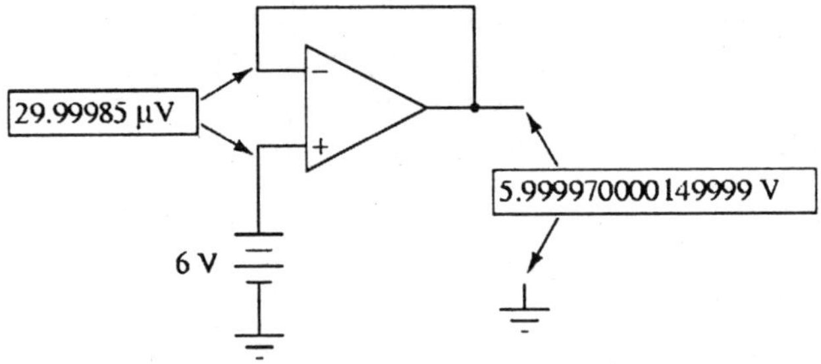

The effects of negative feedback
(rounded figures)

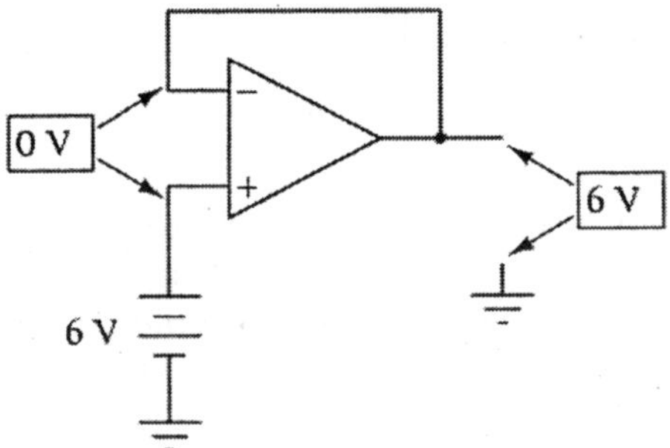

One great advantage to using an op-amp with negative feedback is that the actual voltage gain of the op-amp doesn't matter, so long as its very large. If the op-amp's differential gain were 250,000 instead of 200,000, all it would mean is that the output voltage would hold just a little closer to V_{in} (less differential voltage needed between inputs to generate the required output). In the circuit just illustrated, the output voltage would still be (for all practical purposes) equal to the non-inverting input voltage. Op-amp gains, therefore, do not have to be precisely set by the factory in order for the circuit designer to build an amplifier circuit with precise gain. Negative feedback makes the system self-correcting. The above circuit as a whole will simply follow the input voltage with a stable gain of 1.

Going back to our differential amplifier model, we can think of the operational amplifier as being a variable voltage source controlled by an extremely sensitive *null detector*, the kind of meter movement or other sensitive measurement device used in bridge circuits to detect a condition of balance (zero volts). The "potentiometer" inside the op-amp creating the variable voltage will move to whatever position it must to "balance" the inverting and noninverting input voltages so that the "null detector" has zero voltage across it:

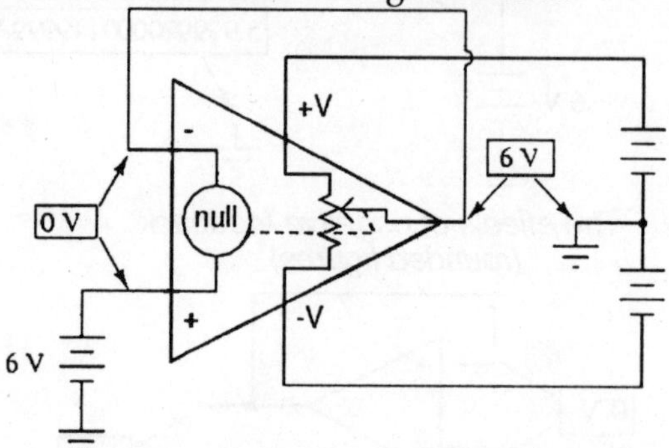

As the "potentiometer" will move to provide an output voltage necessary to satisfy the "null detector" at an "indication" of zero volts, the output voltage becomes equal to the input voltage: in this case, 6 volts. If the input voltage changes at all, the "potentiometer" inside the op-amp will change position to hold the "null detector" in balance (indicating zero volts), resulting in an output voltage approximately equal to the input voltage at all times.

This will hold true within the range of voltages that the op-amp can output. With a power supply of +15V/-15V, and an ideal amplifier that can swing its output voltage just as far, it will faithfully "follow" the input voltage between the limits of +15 volts and -15 volts. For this reason, the above circuit is known as a *voltage follower*. Like its one-transistor counterpart, the common-collector ("emitter-follower") amplifier, it has a voltage gain of 1, a high input impedance, a low output impedance, and a high current gain. Voltage followers are also known as *voltage buffers*, and are used to boost the current-sourcing ability of voltage signals too weak (too high of source impedance) to directly drive a load.

The op-amp model shown in the last illustration depicts how the output voltage is essentially isolated from the input voltage, so that current on the output pin is not supplied by the input voltage source at all, but rather from the power supply powering the op-amp.

It should be mentioned that many op-amps cannot swing their output voltages exactly to +V/-V power supply rail voltages. The model 741 is one of those that cannot: when saturated, its output voltage peaks within about one volt of the +V power supply voltage and within about 2 volts of the -V power supply voltage.

Therefore, with a split power supply of +15/-15 volts, a 741 op-amp's output may go as high as +14 volts or as low as -13 volts (approximately), but no further. This is due to its bipolar transistor design.

These two voltage limits are known as the *positive saturation voltage* and *negative saturation voltage*, respectively. Other op-amps, such as the model 3130 with field-effect transistors in the final output stage, have the ability to swing their output voltages within millivolts of either power supply *rail* voltage. Consequently, their positive and negative saturation voltages are practically equal to the supply voltages.

DIVIDED FEEDBACK

If we add a voltage divider to the negative feedback wiring so that only a *fraction* of the output voltage is fed back to the inverting input instead of the full amount, the output voltage will be a *multiple* of the input voltage (please bear in mind that the power supply connections to the op-amp have been omitted once again for simplicity's sake):

The effects of divided negative feedback

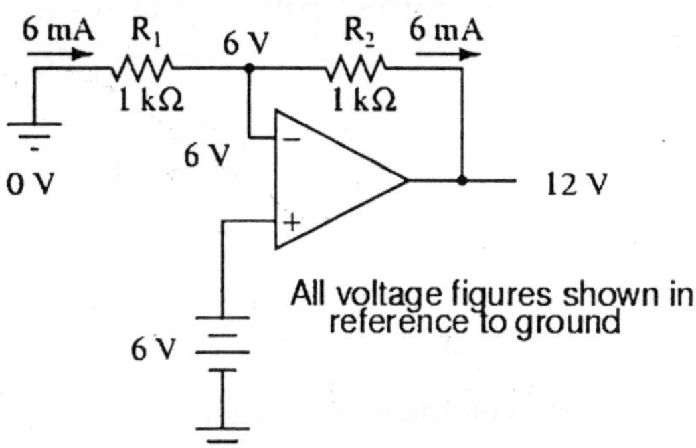

If R_1 and R_2 are both equal and V_{in} is 6 volts, the op-amp will output whatever voltage is needed to drop 6 volts across R_1 (to make the inverting input voltage equal to 6 volts, as well, keeping the voltage difference between the two inputs equal to zero). With the 2:1 voltage divider of R_1 and R_2, this will take 12 volts at the output of the op-amp to accomplish.

Another way of analyzing this circuit is to start by calculating the magnitude and direction of current through R_1, knowing the voltage on either side (and therefore, by subtraction, the voltage across R_1), and R_1's resistance. Since the left-hand side of R_1 is connected to ground (0 volts) and the right-hand side is at a potential of 6 volts (due to the negative feedback holding that point equal to V_{in}), we can see that we have 6 volts across R_1.

This gives us 6 mA of current through R_1 from left to right. Because we know that both inputs of the op-amp have extremely high impedance, we can safely assume they won't add or subtract any current through the divider. In other words, we can treat R_1 and R_2 as being in series with each other:

All of the electrons flowing through R_1 must flow through R_2. Knowing the current through R_2 and the resistance of R_2, we can calculate the voltage across R_2 (6 volts), and its polarity. Counting up voltages from ground (0 volts) to the right-hand side of R_2, we arrive at 12 volts on the output.

Upon examining the last illustration, one might wonder, "where does that 1 mA of current go?" The last illustration doesn't show the entire current path, but in reality it comes from the negative side of the DC power supply, through ground, through R_1, through R_2, through the output pin of the op-amp, and then back to the positive side of the DC power supply through the output transistor(s) of the op-amp. Using the null detector/potentiometer model of the op-amp, the current path looks like this:

The 6 volt signal source does not have to supply any current for the circuit: it merely commands the op-amp to balance voltage between the inverting (-) and noninverting (+)

input pins, and in so doing produce an output voltage that is twice the input due to the dividing effect of the two 1 kΩ resistors.

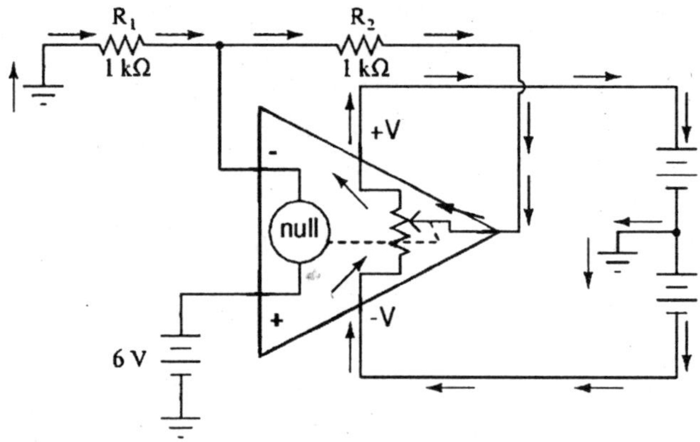

We can change the voltage gain of this circuit, overall, just by adjusting the values of R_1 and R_2 (changing the ratio of output voltage that is fed back to the inverting input). Gain can be calculated by the following formula:

$$A_v \frac{R_2}{R_1} + 1.$$

Note that the voltage gain for this design of amplifier circuit can never be less than 1. If we were to lower R_2 to a value of zero ohms, our circuit would be essentially identical to the voltage follower, with the output directly connected to the inverting input. Since the voltage follower has a gain of 1, this sets the lower gain limit of the noninverting amplifier. However, the gain can be increased far beyond 1, by increasing R_2 in proportion to R_1.

Also note that the polarity of the output matches that of the input, just as with a voltage follower. A positive input voltage results in a positive output voltage, and vice versa (with respect to ground). For this reason, this circuit is referred to as a *noninverting amplifier*.

Just as with the voltage follower, we see that the differential gain of the op-amp is irrelevant, so long as its very high. The voltages and currents in this circuit would hardly change at all if the op-amp's voltage gain were 250,000 instead of 200,000. This stands as a stark contrast to single-transistor amplifier circuit designs, where the Beta of the individual transistor greatly influenced the overall gains of the amplifier. With negative feedback, we have a self-correcting system that amplifies voltage according to the ratios set by the feedback resistors, not the gains internal to the op-amp.

Let's see what happens if we retain negative feedback through a voltage divider, but apply the input voltage at a different location:

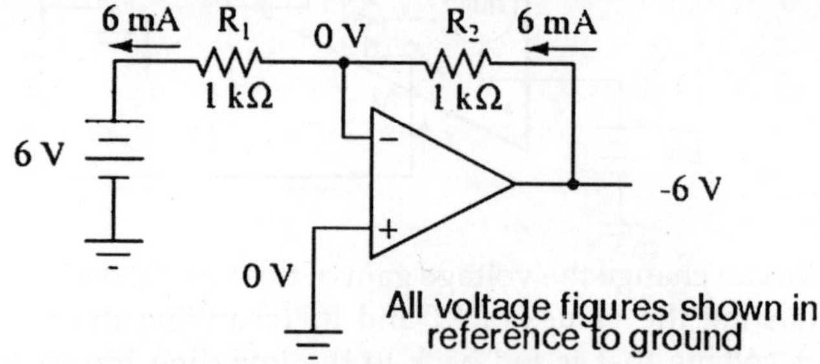

By grounding the noninverting input, the negative feedback from the output seeks to hold the inverting input's voltage at 0 volts, as well. For this reason, the inverting input is referred to in this circuit as a *virtual ground,* being held at ground potential (0 volts) by the feedback, yet not directly connected to (electrically common with) ground. The input voltage this time is applied to the left-hand end of the voltage divider ($R_1 = R_2 = 1$ kΩ again), so the output voltage must swing to -6 volts in order to balance the middle at ground potential (0 volts).

Using the same techniques as with the noninverting amplifier, we can analyze this circuit's operation by determining current magnitudes and directions, starting with R_1, and continuing on to determining the output voltage.

We can change the overall voltage gain of this circuit, overall, just by adjusting the values of R_1 and R_2 (changing the ratio of output voltage that is fed back to the inverting input). Gain can be calculated by the following formula:

$$A_V \frac{R_2}{R_1}$$

Note that this circuit's voltage gain *can* be less than 1, depending solely on the ratio of R_2 to R_1. Also note that the output voltage is always the opposite polarity of the input voltage.

A positive input voltage results in a negative output voltage, and vice versa (with respect to ground). For this reason, this circuit is referred to as an *inverting amplifier*. Sometimes, the gain formula contains a negative sign (before the R_2/R_1 fraction) to reflect this reversal of polarities.

These two amplifier circuits we've just investigated serve the purpose of multiplying or dividing the magnitude of the input voltage signal. This is exactly how the mathematical operations of multiplication and division are typically handled in analog computer circuitry.

AN ANALOGY FOR DIVIDED FEEDBACK

A helpful analogy for understanding divided feedback amplifier circuits is that of a mechanical lever, with relative motion of the lever's ends representing change in input and output voltages, and the fulcrum (pivot point) representing the location of the ground point, real or virtual.

Take for example the following non inverting op-amp circuit. We know from the prior section that the voltage gain of a non inverting amplifier configuration can never be less than unity (1).

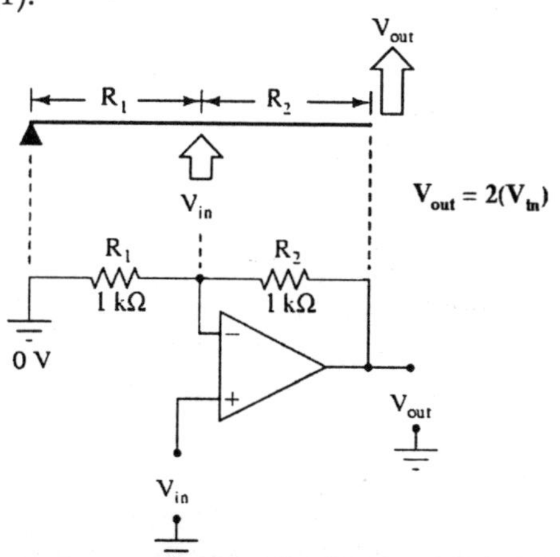

If we draw a lever diagram next to the amplifier schematic, with the distance between fulcrum and lever ends representative of resistor values, the motion of the lever will signify changes in voltage at the input and output terminals of the amplifier:

Physicists call this type of lever, with the input force (effort) applied between the fulcrum and output (load), a *third-class* lever.

It is characterized by an output displacement (motion) at least as large than the input displacement – a "gain" of at least 1 – and in the same direction. Applying a positive input voltage to this op-amp circuit is analogous to displacing the "input" point on the lever upward:

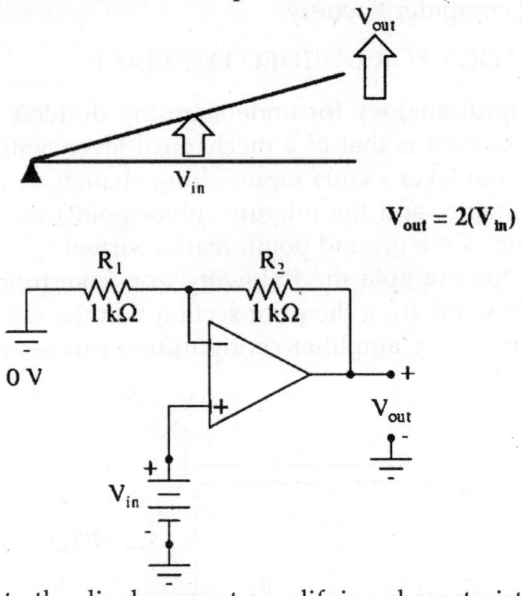

Due to the displacement-amplifying characteristics of the lever, the "output" point will move twice as far as the "input" point, and in the same direction. In the electronic circuit, the output voltage will equal twice the input, with the same polarity.

Applying a negative input voltage is analogous to moving the lever downward from its level "zero" position, resulting in an amplified output displacement that is also negative:

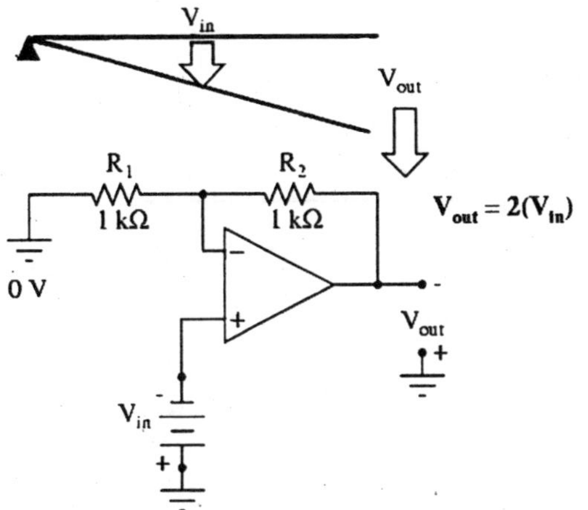

If we alter the resistor ratio R_2/R_1, we change the gain of the op-amp circuit. In lever terms, this means moving the input point in relation to the fulcrum and lever end, which similarly changes the displacement "gain" of the machine:

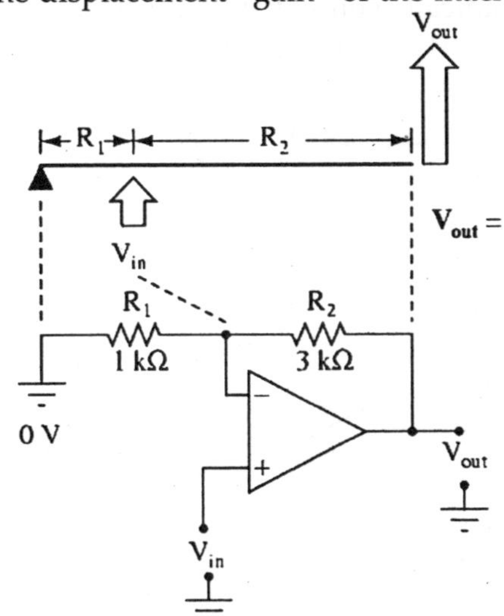

Now, any input signal will become amplified by a factor of four instead of by a factor of two:

Inverting op-amp circuits may be modeled using the lever analogy as well. With the inverting configuration, the ground point of the feedback voltage divider is the op-amp's inverting input with the input to the left and the output to the right.

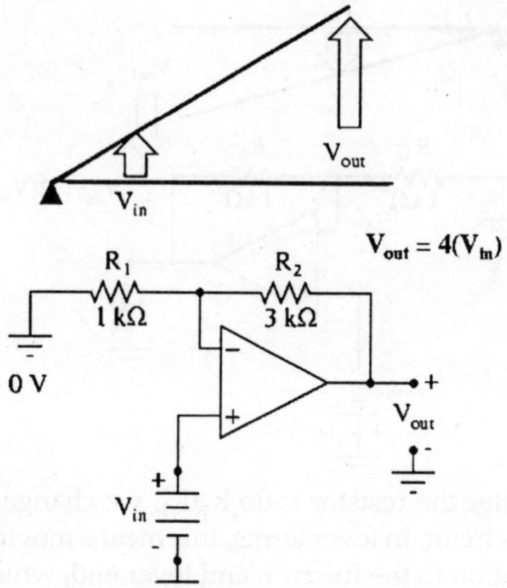

This is mechanically equivalent to a *first-class* lever, where the input force (effort) is on the opposite side of the fulcrum from the output (load):

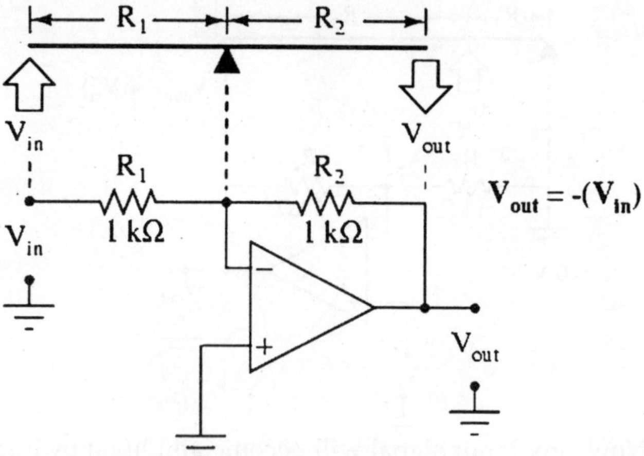

With equal-value resistors (equal-lengths of lever on each side of the fulcrum), the output voltage (displacement) will be equal in magnitude to the input voltage (displacement), but of the opposite polarity (direction). A positive input results in a negative output:

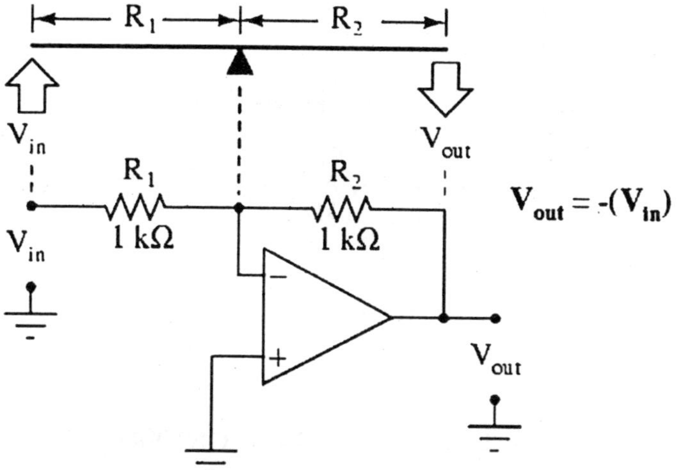

Changing the resistor ratio R_2/R_1 changes the gain of the amplifier circuit, just as changing the fulcrum position on the lever changes its mechanical displacement "gain." Consider the following example, where R_2 is made twice as large as R_1:

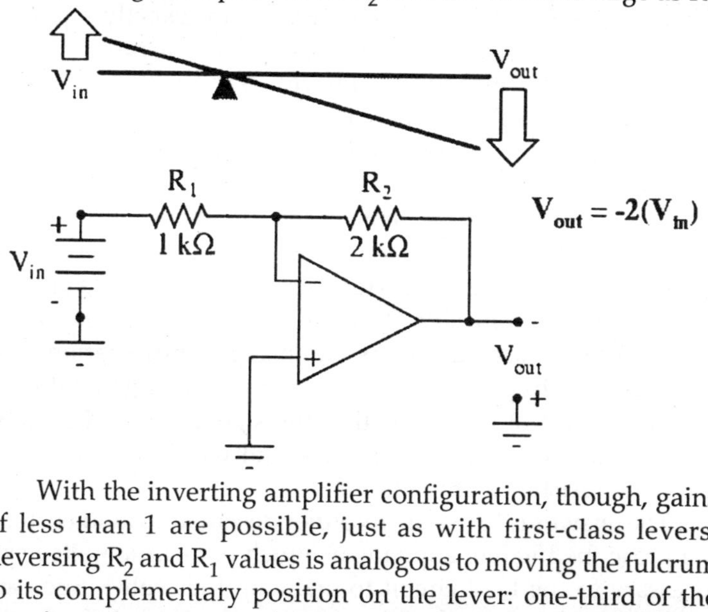

With the inverting amplifier configuration, though, gains of less than 1 are possible, just as with first-class levers. Reversing R_2 and R_1 values is analogous to moving the fulcrum to its complementary position on the lever: one-third of the way from the output end. There, the output displacement will be one-half the input displacement:

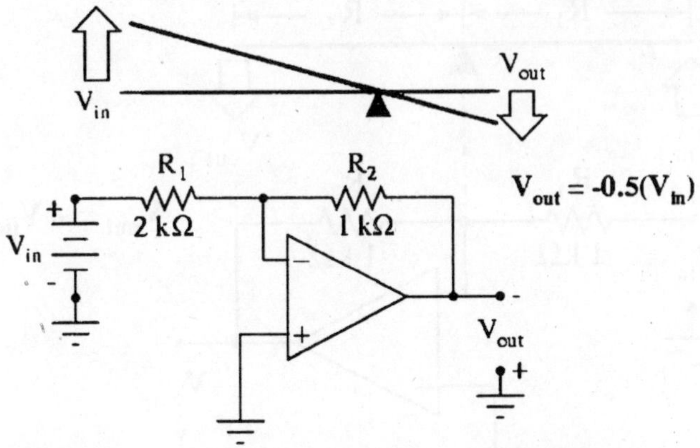

VOLTAGE-TO-CURRENT SIGNAL CONVERSION

In instrumentation circuitry, DC signals are often used as analog representations of physical measurements such as temperature, pressure, flow, weight, and motion. Most commonly, *DC current* signals are used in preference to *DC voltage* signals, because current signals are exactly equal in magnitude throughout the series circuit loop carrying current from the source (measuring device) to the load (indicator, recorder, or controller), whereas voltage signals in a parallel circuit may vary from one end to the other due to resistive wire losses.

Furthermore, current-sensing instruments typically have low impedances (while voltage-sensing instruments have high impedances), which gives current-sensing instruments greater electrical noise immunity.

In order to use current as an analog representation of a physical quantity, we have to have some way of generating a precise amount of current within the signal circuit. But how do we generate a precise current signal when we might not know the resistance of the loop?

The answer is to use an amplifier designed to hold current to a prescribed value, applying as much or as little voltage as necessary to the load circuit to maintain that value. Such an amplifier performs the function of a *current source*. An op-amp with negative feedback is a perfect candidate for such a task:

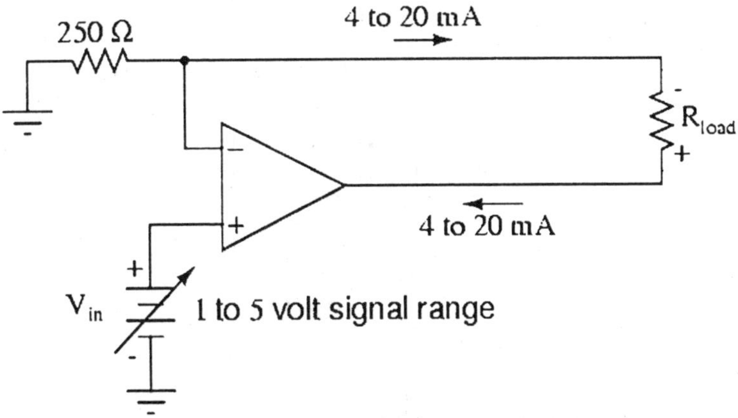

The input voltage to this circuit is assumed to be coming from some type of physical transducer/amplifier arrangement, calibrated to produce 1 volt at 0 percent of physical measurement, and 5 volts at 100 percent of physical measurement. The standard analog current signal range is 4 mA to 20 mA, signifying 0% to 100% of measurement range, respectively. At 5 volts input, the 250 Ω (precision) resistor will have 5 volts applied across it, resulting in 20 mA of current in the large loop circuit (with R_{load}). It does not matter what resistance value R_{load} is, or how much wire resistance is present in that large loop, so long as the op-amp has a high enough power supply voltage to output the voltage necessary to get 20 mA flowing through R_{load}. The 250 Ω resistor establishes the relationship between input voltage and output current, in this case creating the equivalence of 1-5 V in/4-20 mA out. If we were converting the 1-5 volt input signal to a 10-50 mA output signal (an older, obsolete instrumentation standard for industry), we'd use a 100 Ω precision resistor instead.

Another name for this circuit is *transconductance amplifier*. In electronics, transconductance is the mathematical ratio of current change divided by voltage change ($\Delta I / \Delta V$), and it is measured in the unit of Siemens, the same unit used to express conductance (the mathematical reciprocal of resistance: current/voltage). In this circuit, the transconductance ratio is fixed by the value of the 250 Ω resistor, giving a linear current-out/voltage-in relationship.

AVERAGER AND SUMMER CIRCUITS

If we take three equal resistors and connect one end of each to a common point, then apply three input voltages (one to each of the resistors' free ends), the voltage seen at the common point will be the mathematical *average* of the three.

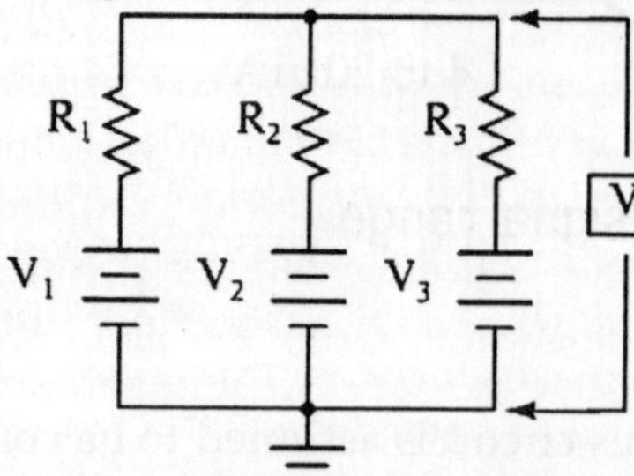

$$V_{out} = \frac{\frac{V_1}{R_1} + \frac{V_2}{R_2} + \frac{V_3}{R_3}}{\frac{1}{R_1} + \frac{1}{R_2} + \frac{1}{R_3}}$$

This circuit is really nothing more than a practical application of Millman's Theorem:

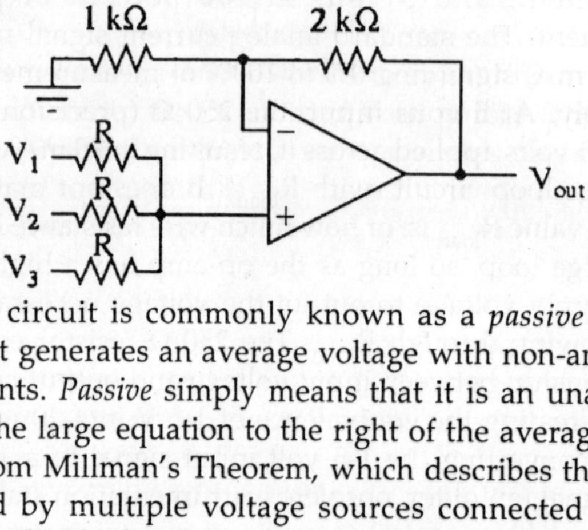

This circuit is commonly known as a *passive averager*, because it generates an average voltage with non-amplifying components. *Passive* simply means that it is an unamplified circuit. The large equation to the right of the averager circuit comes from Millman's Theorem, which describes the voltage produced by multiple voltage sources connected together through individual resistances. Since the three resistors in the averager circuit are equal to each other, we can simplify Millman's formula by writing R_1, R_2, and R_3 simply as R (one, equal resistance instead of three individual resistances):

$$V_{out} = \frac{\frac{V_1}{R} + \frac{V_2}{R} + \frac{V_3}{R}}{\frac{1}{R} + \frac{1}{R} + \frac{1}{R}}$$

$$V_{out} = \frac{\dfrac{V_1 + V_2 + V_3}{R}}{\dfrac{3}{R}}$$

$$V_{out} = \frac{V_1 + V_2 + V_3}{3}.$$

If we take a passive averager and use it to connect three input voltages into an op-amp amplifier circuit with a gain of 3, we can turn this *averaging* function into an *addition* function. The result is called a *noninverting summer* circuit:

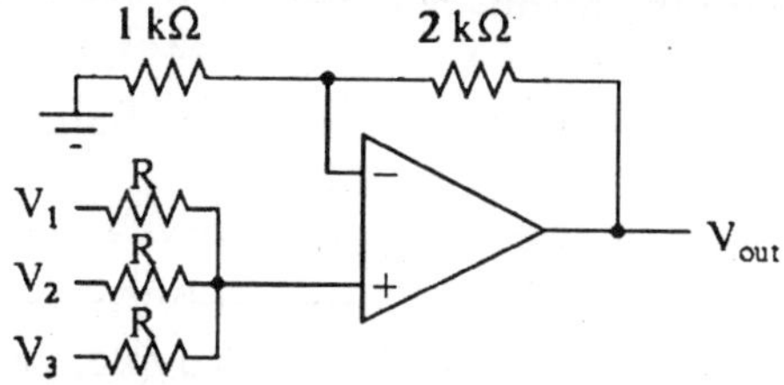

With a voltage divider composed of a 2 kΩ/1 kΩ combination, the noninverting amplifier circuit will have a voltage gain of 3. By taking the voltage from the passive averager, which is the sum of V_1, V_2, and V_3 divided by 3, and multiplying that average by 3, we arrive at an output voltage equal to the *sum* of V_1, V_2, and V_3:

$$V_{out} = 3\,\frac{V_1 + V_2 + V_3}{3}$$

$$V_{out} = V_1 + V_2 + V_3.$$

Much the same can be done with an inverting op-amp amplifier, using a passive averager as part of the voltage divider feedback circuit. The result is called an *inverting summer* circuit:

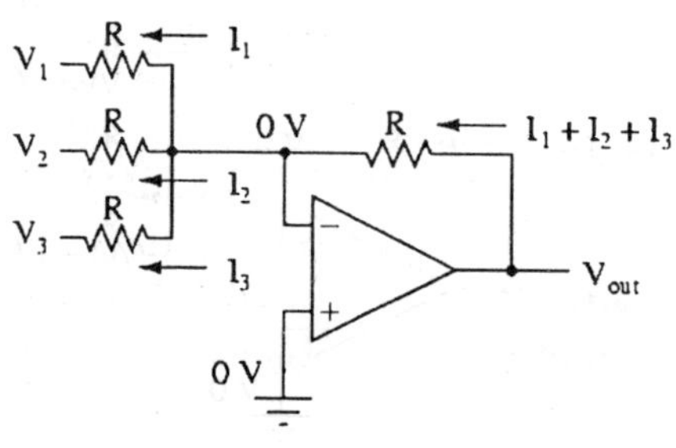

Now, with the right-hand sides of the three averaging resistors connected to the virtual ground point of the op-amp's inverting input, Millman's Theorem no longer directly applies as it did before. The voltage at the virtual ground is now held at 0 volts by the op-amp's negative feedback, whereas before it was free to float to the average value of V_1, V_2, and V_3. However, with all resistor values equal to each other, the currents through each of the three resistors will be proportional to their respective input voltages. Since those three currents will *add* at the virtual ground node, the algebraic sum of those currents through the feedback resistor will produce a voltage at V_{out} equal to $V_1 + V_2 + V_3$, except with reversed polarity. The reversal in polarity is what makes this circuit an *inverting* summer:

$$V_{out} = -(V_1 + V_2 + V_3).$$

Summer (adder) circuits are quite useful in analog computer design, just as multiplier and divider circuits would be. Again, it is the extremely high differential gain of the op-amp which allows us to build these useful circuits with a bare minimum of components.

BUILDING A DIFFERENTIAL AMPLIFIER

An op-amp with no feedback is already a differential amplifier, amplifying the voltage difference between the two inputs. However, its gain cannot be controlled, and it is generally too high to be of any practical use.

So far, our application of negative feedback to op-amps has resulting in the practical loss of one of the inputs, the resulting amplifier only good for amplifying a single voltage signal input. With a little ingenuity, however, we can construct an op-amp circuit maintaining both voltage inputs, yet with a controlled gain set by external resistors.

If all the resistor values are equal, this amplifier will have a differential voltage gain of 1. The analysis of this circuit is essentially the same as that of an inverting amplifier, except that the noninverting input (+) of the op-amp is at a voltage equal to a fraction of V_2, rather than being connected directly to ground.

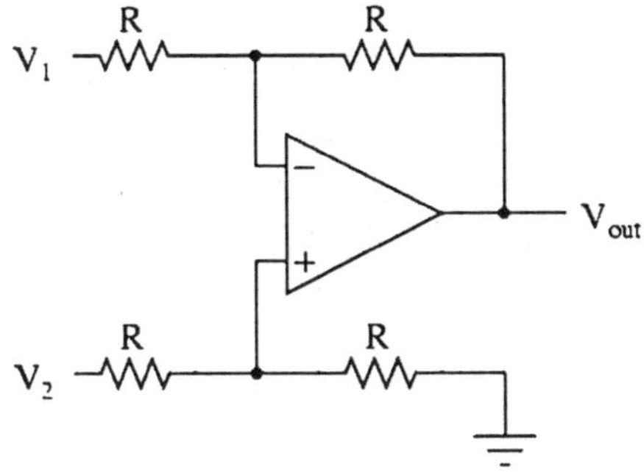

$$V_{out} = V_2 - V_1.$$

As would stand to reason, V_2 functions as the noninverting input and V_1 functions as the inverting input of the final amplifier circuit. Therefore:

If we wanted to provide a differential gain of anything other than 1, we would have to adjust the resistances in *both* upper and lower voltage dividers, necessitating multiple resistor changes and balancing between the two dividers for symmetrical operation. This is not always practical, for obvious reasons.

Another limitation of this amplifier design is the fact that its input impedances are rather low compared to that of some other op-amp configurations, most notably the noninverting (single-ended input) amplifier. Each input voltage source has to drive current through a resistance, which constitutes far less impedance than the bare input of an op-amp alone. The solution to this problem, fortunately, is quite simple. All we need to do is "buffer" each input voltage signal through a voltage follower like this:

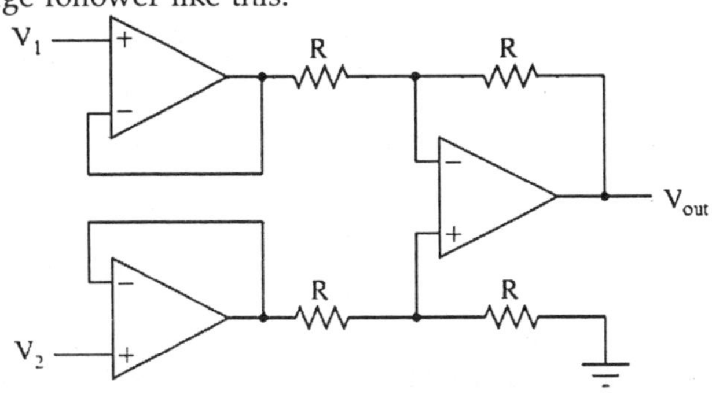

Now the V_1 and V_2 input lines are connected straight to the inputs of two voltage-follower op-amps, giving very high impedance. The two op-amps on the left now handle the driving of current through the resistors instead of letting the input voltage sources (whatever they may be) do it. The increased complexity to our circuit is minimal for a substantial benefit.

THE INSTRUMENTATION AMPLIFIER

As suggested before, it is beneficial to be able to adjust the gain of the amplifier circuit without having to change more than one resistor value, as is necessary with the previous design of differential amplifier. The so-called *instrumentation* builds on the last version of differential amplifier to give us that capability:

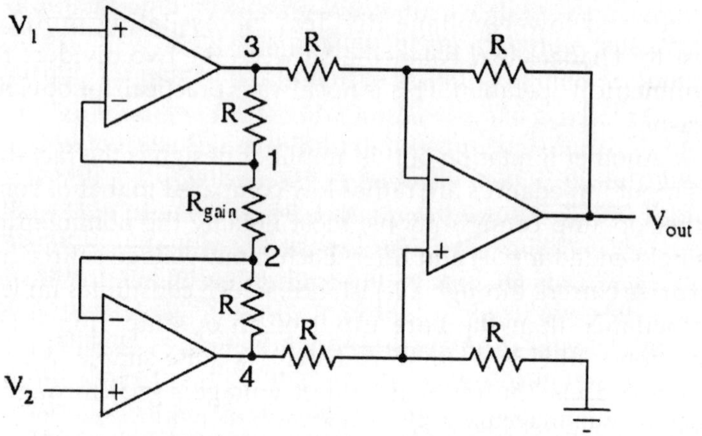

This intimidating circuit is constructed from a buffered differential amplifier stage with three new resistors linking the two buffer circuits together. Consider all resistors to be of equal value except for R_{gain}. The negative feedback of the upper-left op-amp causes the voltage at point 1 (top of R_{gain}) to be equal to V_1. Likewise, the voltage at point 2 (bottom of R_{gain}) is held to a value equal to V_2. This establishes a voltage drop across R_{gain} equal to the voltage difference between V_1 and V_2. That voltage drop causes a current through R_{gain}, and since the feedback loops of the two input op-amps draw no current, that same amount of current through R_{gain} must be going through the two "R" resistors above and below it. This produces a voltage drop between points 3 and 4 equal to:

$$V_{3-4} = (V_2 - V_1)\left(1+\frac{2R}{R_{gain}}\right).$$

The regular differential amplifier on the right-hand side of the circuit then takes this voltage drop between points 3 and 4, and amplifies it by a gain of 1 (assuming again that all "R" resistors are of equal value). Though this looks like a cumbersome way to build a differential amplifier, it has the distinct advantages of possessing extremely high input impedances on the V_1 and V_2 inputs (because they connect straight into the noninverting inputs of their respective op-amps), and adjustable gain that can be set by a single resistor. Manipulating the above formula a bit, we have a general expression for overall voltage gain in the instrumentation amplifier:

$$A_V = \left(1+\frac{2R}{R_{gain}}\right).$$

Though it may not be obvious by looking at the schematic, we can change the differential gain of the instrumentation amplifier simply by changing the value of one resistor: R_{gain}. Yes, we could still change the overall gain by changing the values of some of the other resistors, but this would necessitate *balanced* resistor value changes for the circuit to remain symmetrical. Please note that the lowest gain possible with the above circuit is obtained with R_{gain} completely open (infinite resistance), and that gain value is 1.

DIFFERENTIATOR AND INTEGRATOR CIRCUITS

By introducing electrical reactance into the feedback loops of op-amp amplifier circuits, we can cause the output to respond to changes in the input voltage over *time*. Drawing their names from their respective calculus functions, the *integrator* produces a voltage output proportional to the product (multiplication) of the input voltage and time; and the *differentiator* (not to be confused with *differential*) produces a voltage output proportional to the input voltage's rate of change.

Capacitance can be defined as the measure of a capacitor's opposition to changes in voltage. The greater the capacitance, the more the opposition. Capacitors oppose voltage change by creating current in the circuit: that is, they either charge or discharge in response to a change in applied voltage. So, the more capacitance a capacitor has, the greater its charge or discharge current will be for any given rate of voltage change across it. The equation for this is quite simple:

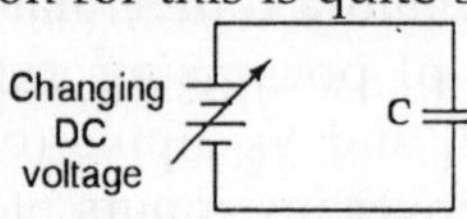

$$i = C \frac{dv}{dt}$$

The *dv/dt* fraction is a calculus expression representing the rate of voltage change over time. If the DC supply in the above circuit were steadily increased from a voltage of 15 volts to a voltage of 16 volts over a time span of 1 hour, the current through the capacitor would most likely be *very* small, because of the very low rate of voltage change (dv/dt = 1 volt/3600 seconds). However, if we steadily increased the DC supply from 15 volts to 16 volts over a shorter time span of 1 second, the rate of voltage change would be much higher, and thus the charging current would be much higher (3600 times higher, to be exact). Same amount of change in voltage, but vastly different *rates* of change, resulting in vastly different amounts of current in the circuit.

To put some definite numbers to this formula, if the voltage across a 47 μF capacitor was changing at a linear rate of 3 volts per second, the current "through" the capacitor would be (47 μF)(3 V/s) = 141 μA.

We can build an op-amp circuit which measures change in voltage by measuring current through a capacitor, and outputs a voltage proportional to that current:

The right-hand side of the capacitor is held to a voltage of 0 volts, due to the "virtual ground" effect. Therefore, current "through" the capacitor is solely due to *change* in the input voltage. A steady input voltage won't cause a current through C, but a *changing* input voltage will.

Differentiator

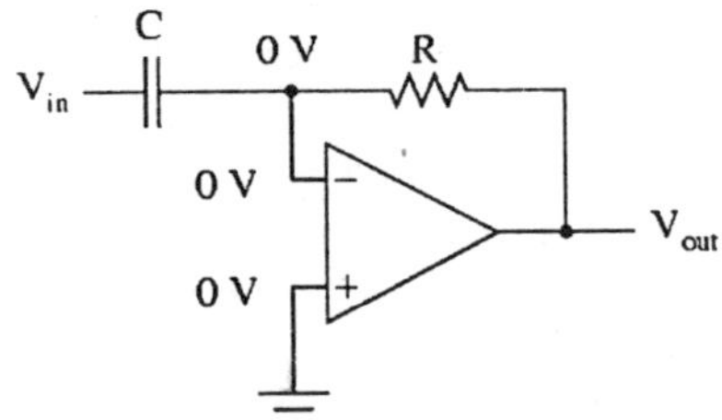

Capacitor current moves through the feedback resistor, producing a drop across it, which is the same as the output voltage. A linear, positive rate of input voltage change will result in a steady negative voltage at the output of the op-amp. Conversely, a linear, negative rate of input voltage change will result in a steady positive voltage at the output of the op-amp. This polarity inversion from input to output is due to the fact that the input signal is being sent (essentially) to the inverting input of the op-amp, so it acts like the inverting amplifier mentioned previously. The faster the rate of voltage change at the input (either positive or negative), the greater the voltage at the output.

The formula for determining voltage output for the differentiator is as follows:

$$V_{out} = -\ RC\frac{d\,v_{in}}{dt}\ .$$

Applications for this, besides representing the derivative calculus function inside of an analog computer, include rate-of-change indicators for process instrumentation. One such rate-of-change signal application might be for monitoring (or controlling) the rate of temperature change in a furnace, where too high or too low of a temperature rise rate could be detrimental. The DC voltage produced by the differentiator circuit could be used to drive a comparator, which would signal an alarm or activate a control if the rate of change exceeded a pre-set level.

In process control, the derivative function is used to make control decisions for maintaining a process at setpoint, by monitoring the rate of process change over time and taking action to prevent excessive rates of change, which can lead to an unstable condition. Analog electronic controllers use variations of this circuitry to perform the derivative function.

On the other hand, there are applications where we need precisely the opposite function, called *integration* in calculus. Here, the op-amp circuit would generate an output voltage proportional to the magnitude and duration that an input voltage signal has deviated from 0 volts. Stated differently, a constant input signal would generate a certain *rate of change* in the output voltage: differentiation in reverse. To do this, all we have to do is swap the capacitor and resistor in the previous circuit:

Integrator

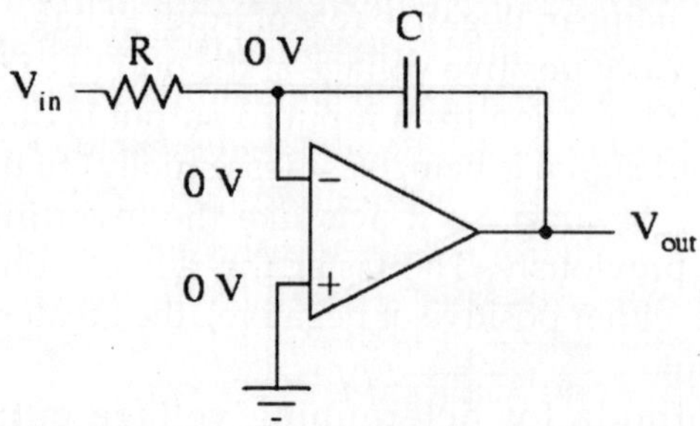

As before, the negative feedback of the op-amp ensures that the inverting input will be held at 0 volts (the virtual ground). If the input voltage is exactly 0 volts, there will be no current through the resistor, therefore no charging of the capacitor, and therefore the output voltage will not change. We cannot guarantee what voltage will be at the output with respect to ground in this condition, but we can say that the output voltage *will be constant*.

However, if we apply a constant, positive voltage to the input, the op-amp output will fall negative at a linear rate, in an attempt to produce the changing voltage across the capacitor necessary to maintain the current established by the voltage difference across the resistor. Conversely, a constant, negative voltage at the input results in a linear, rising (positive) voltage at the output. The output voltage rate-of-change will be proportional to the value of the input voltage.

The formula for determining voltage output for the integrator is as follows:

$$\frac{d\text{v}_{out}}{dt} = -\frac{V_{in}}{RC}$$

or

$$V_{out} = \int_0^t -\frac{V_{in}}{RC}\,dt + C$$

Where,

c = Output voltage at start time (t = 0)

One application for this device would be to keep a "running total" of radiation exposure, or dosage, if the input voltage was a proportional signal supplied by an electronic radiation detector. Nuclear radiation can be just as damaging at low intensities for long periods of time as it is at high intensities for short periods of time. An integrator circuit would take both the intensity (input voltage magnitude) and time into account, generating an output voltage representing total radiation dosage.

Another application would be to integrate a signal representing water flow, producing a signal representing total quantity of water that has passed by the flow meter. This application of an integrator is sometimes called a *totalizer* in the industrial instrumentation trade.

POSITIVE FEEDBACK

As we've seen, negative feedback is an incredibly useful principle when applied to operational amplifiers. It is what allows us to create all these practical circuits, being able to precisely set gains, rates, and other significant parameters with just a few changes of resistor values. Negative feedback makes all these circuits stable and self-correcting.

The basic principle of negative feedback is that the output tends to drive in a direction that creates a condition of equilibrium (balance). In an op-amp circuit with no feedback, there is no corrective mechanism, and the output voltage will saturate with the tiniest amount of differential voltage applied between the inputs. The result is a comparator:

With negative feedback (the output voltage "fed back" somehow to the inverting input), the circuit tends to prevent itself from driving the output to full saturation. Rather, the output voltage drives only as high or as low as needed to balance the two inputs' voltages:

Negative feedback

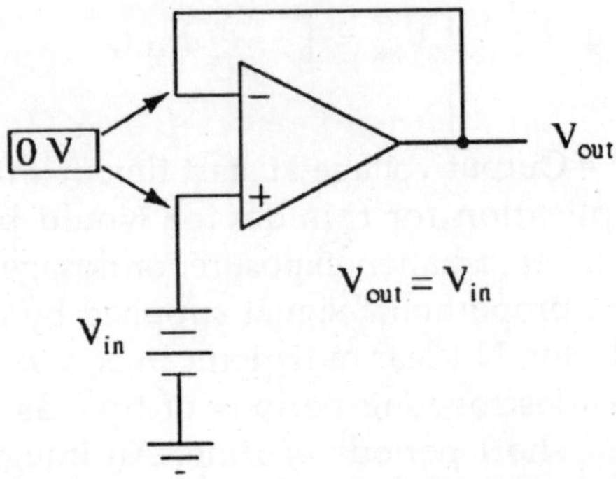

Whether the output is directly fed back to the inverting (–) input or coupled through a set of components, the effect is the same: the extremely high differential voltage gain of the op-amp will be "tamed" and the circuit will respond according to the dictates of the feedback "loop" connecting output to inverting input.

Another type of feedback, namely *positive feedback*, also finds application in op-amp circuits. Unlike negative feedback, where the output voltage is "fed back" to the inverting (–) input, with positive feedback the output voltage is somehow routed back to the noninverting (+) input. In its simplest form, we could connect a straight piece of wire from output to noninverting input and see what happens:

Positive feedback

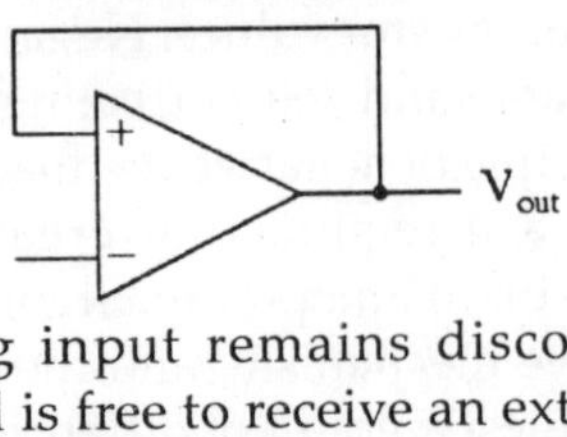

The inverting input remains disconnected from the feedback loop, and is free to receive an external voltage. Let's see what happens if we ground the inverting input:

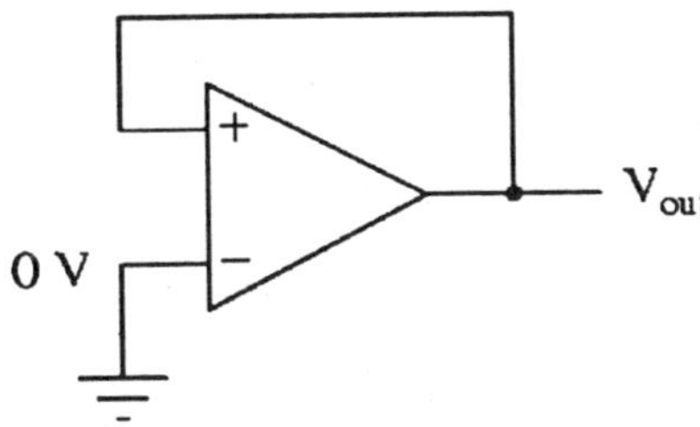

With the inverting input grounded (maintained at zero volts), the output voltage will be dictated by the magnitude and polarity of the voltage at the noninverting input. If that voltage happens to be positive, the op-amp will drive its output positive as well, feeding that positive voltage back to the noninverting input, which will result in full positive output saturation.

On the other hand, if the voltage on the noninverting input happens to start out negative, the op-amp's output will drive in the negative direction, feeding back to the noninverting input and resulting in full negative saturation.

What we have here is a circuit whose output is *bistable*: stable in one of two states (saturated positive or saturated negative). Once it has reached one of those saturated states, it will tend to remain in that state, unchanging. What is necessary to get it to switch states is a voltage placed upon the inverting (-) input of the same polarity, but of a slightly greater magnitude. For example, if our circuit is saturated at an output voltage of +12 volts, it will take an input voltage at the inverting input of at least +12 volts to get the output to change. When it changes, it will saturate fully negative.

So, an op-amp with positive feedback tends to stay in whatever output state its already in. It "latches" between one of two states, saturated positive or saturated negative. Technically, this is known as *hysteresis*.

Hysteresis can be a useful property for a comparator circuit to have. As we've seen before, comparators can be used to produce a square wave from any sort of ramping waveform (sine wave, triangle wave, saw tooth wave, etc.) input. If the incoming AC waveform is noise-free (that is, a "pure" waveform), a simple comparator will work just fine.

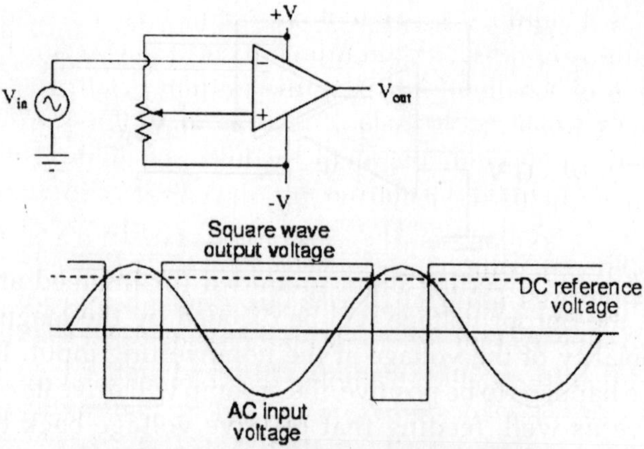

A "clean" AC input waveform produces predictable transition points on the output voltage square wave

However, if there exist any anomalies in the waveform such as harmonics or "spikes" which cause the voltage to rise and fall significantly within the timespan of a single cycle, a comparator's output might switch states unexpectedly:

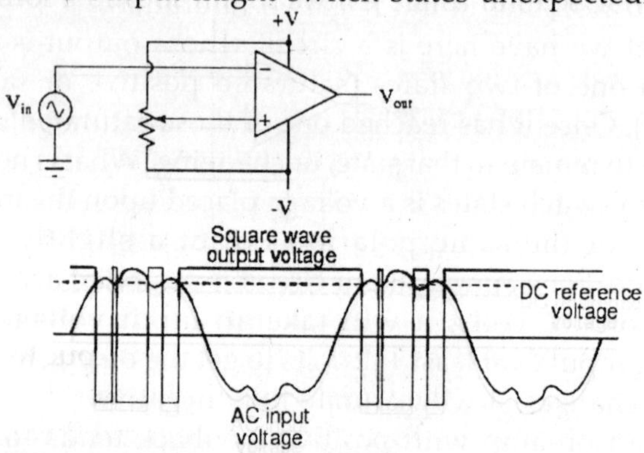

Any time there is a transition through the reference voltage level, no matter how tiny that transition may be, the output of the comparator will switch states, producing a square wave with "glitches."

If we add a little positive feedback to the comparator circuit, we will introduce hysteresis into the output. This hysteresis will cause the output to remain in its current state unless the AC input voltage undergoes a *major* change in magnitude.

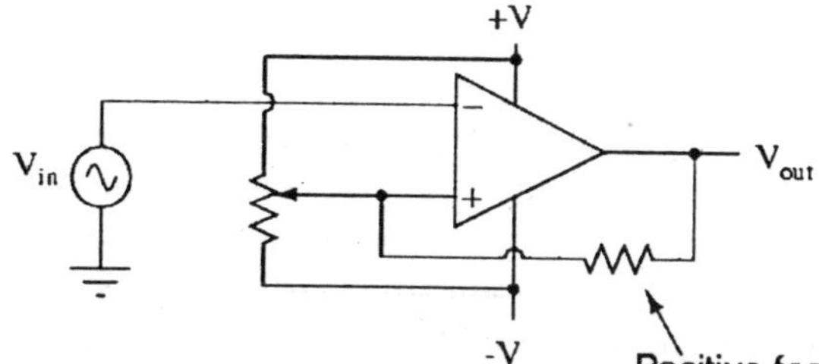

What this feedback resistor creates is a dual-reference for the comparator circuit. The voltage applied to the noninverting (+) input as a reference which to compare with the incoming AC voltage changes depending on the value of the op-amp's output voltage. When the op-amp output is saturated positive, the reference voltage at the noninverting input will be more positive than before. Conversely, when the op-amp output is saturated negative, the reference voltage at the noninverting input will be more negative than before. The result is easier to understand on a graph:

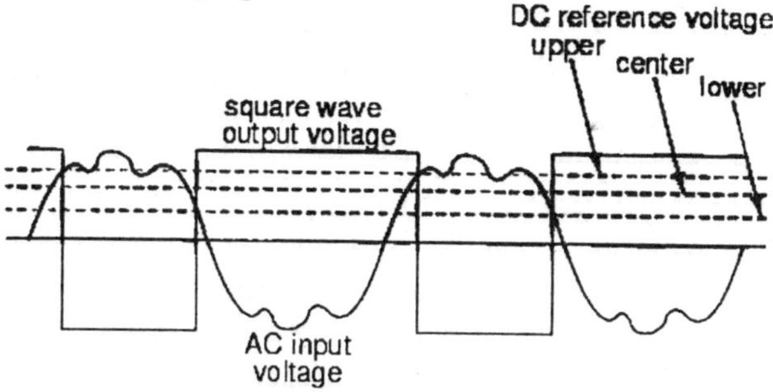

When the op-amp output is saturated positive, the upper reference voltage is in effect, and the output won't drop to a negative saturation level unless the AC input rises *above* that upper reference level. Conversely, when the op-amp output is saturated negative, the lower reference voltage is in effect, and the output won't rise to a positive saturation level unless the AC input drops *below* that lower reference level. The result is a clean square-wave output again, despite significant amounts of distortion in the AC input signal. In order for a "glitch" to cause the comparator to switch from one state to another, it would have to be at least as big (tall) as the difference between the upper and lower reference voltage levels, and at the right point in time to cross both those levels.

Another application of positive feedback in op-amp circuits is in the construction of oscillator circuits. An *oscillator* is a device that produces an alternating (AC), or at least pulsing, output voltage. Technically, it is known as an *astable* device: having no stable output state (no equilibrium whatsoever). Oscillators are very useful devices, and they are easily made with just an op-amp and a few external components.

Oscillator circuit using positive feedback

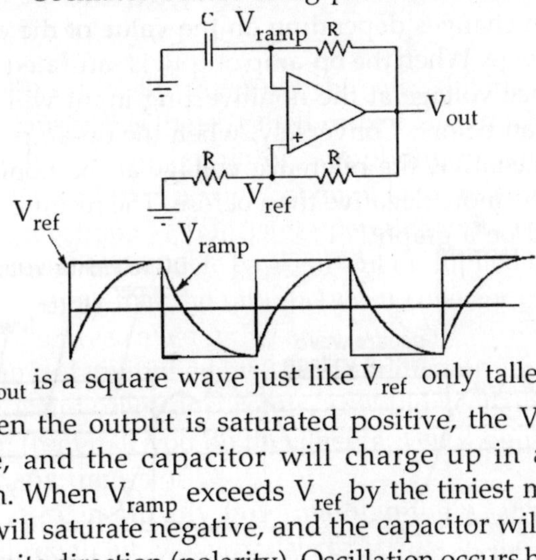

V_{out} is a square wave just like V_{ref} only taller

When the output is saturated positive, the V_{ref} will be positive, and the capacitor will charge up in a positive direction. When V_{ramp} exceeds V_{ref} by the tiniest margin, the output will saturate negative, and the capacitor will charge in the opposite direction (polarity). Oscillation occurs because the positive feedback is instantaneous and the negative feedback is delayed (by means of an RC time constant). The frequency of this oscillator may be adjusted by varying the size of any component.

PRACTICAL CONSIDERATIONS

Real operational have some imperfections compared to an "ideal" model. A real device deviates from a perfect difference amplifier. One minus one may not be zero. It may have have an offset like an analog meter which is not zeroed. The inputs may draw current. The characteristics may drift with age and temperature. Gain may be reduced at high frequencies, and phase may shift from input to output. These imperfection may cause no noticable errors in some applications, unacceptable errors in others. In some cases these errors may be compensated for. Sometimes a higher quality, higher cost device is required.

Common-mode Gain

As stated before, an ideal differential amplifier only amplifies the voltage *difference* between its two inputs. If the two inputs of a differential amplifier were to be shorted together (thus ensuring zero potential difference between them), there should be no change in output voltage for any amount of voltage applied between those two shorted inputs and ground:

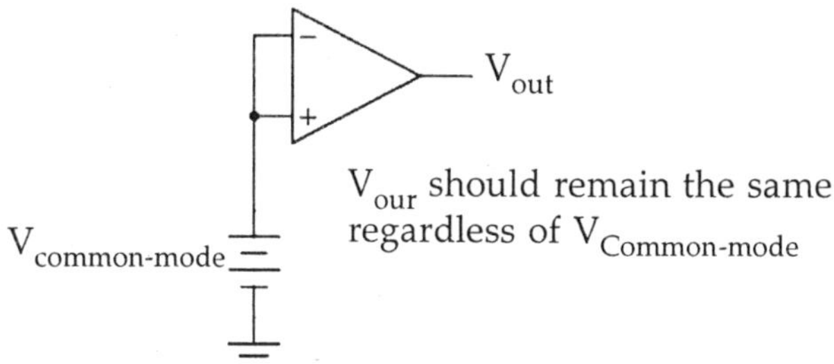

Voltage that is common between either of the inputs and ground, as "$V_{common\text{-}mode}$" is in this case, is called *common-mode voltage*. As we vary this common voltage, the perfect differential amplifier's output voltage should hold absolutely steady (no change in output for any arbitrary change in common-mode input). This translates to a *common-mode voltage gain* of zero.

$$A_V = \frac{\text{Change in } V_{out}}{\text{Change in } V_{in}}$$

. . . if change in $V_{out} = 0$ . . .

$$\frac{0}{\text{Change in } V_{in}} = 0$$

$$A_V = 0$$

The operational amplifier, being a differential amplifier with high differential gain, would ideally have zero common-mode gain as well. In real life, however, this is not easily attained. Thus, common-mode voltages will invariably have some effect on the op-amp's output voltage.

The performance of a real op-amp in this regard is most commonly measured in terms of its differential voltage gain (how much it amplifies the difference between two input voltages) versus its common-mode voltage gain (how much it amplifies a common-mode voltage). The ratio of the former to the latter is called the *common-mode rejection ratio*, abbreviated as CMRR:

$$\text{CMRR} = \frac{\text{Differential}}{\text{Common-mode } A_v}.$$

An ideal op-amp, with zero common-mode gain would have an infinite CMRR. Real op-amps have high CMRRs, the ubiquitous 741 having something around 70 dB, which works out to a little over 3,000 in terms of a ratio.

Because the common mode rejection ratio in a typical op-amp is so high, common-mode gain is usually not a great concern in circuits where the op-amp is being used with negative feedback. If the common-mode input voltage of an amplifier circuit were to suddenly change, thus producing a corresponding change in the output due to common-mode gain, that change in output would be quickly corrected as negative feedback and differential gain (being *much* greater than common-mode gain) worked to bring the system back to equilibrium. Sure enough, a change might be seen at the output, but it would be a lot smaller than what you might expect.

A consideration to keep in mind, though, is common-mode gain in differential op-amp circuits such as instrumentation amplifiers. Outside of the op-amp's sealed package and extremely high differential gain, we may find common-mode gain introduced by an imbalance of resistor values. To demonstrate this, analysis on an instrumentation amplifier with inputs shorted together (no differential voltage), imposing a common-mode voltage to see what happens. First, we'll run the analysis showing the output voltage of a perfectly balanced circuit. We should expect to see no change in output voltage as the common-mode voltage changes:

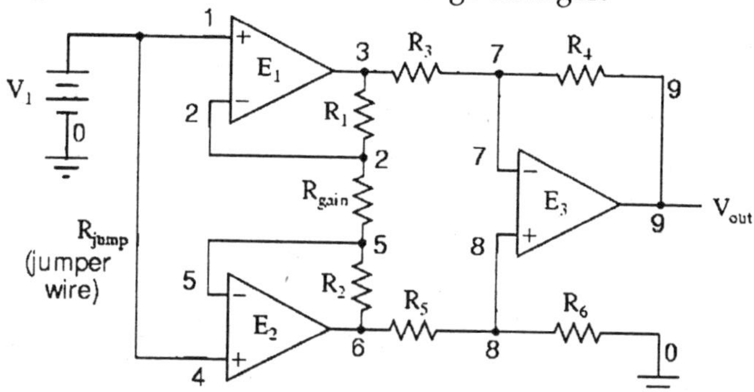

Aside from very small deviations (actually due to quirks rather than real behavior of the circuit), the output remains stable where it should be: at 0 volts, with zero input voltage differential. However, let's introduce a resistor imbalance in the circuit, increasing the value of R_5 from 10,000 Ω to 10,500 Ω, and see what happens (the netlist has been omitted for brevity — the only thing altered is the value of R_5):

Our input voltage differential is still zero volts, yet the output voltage changes significantly as the common-mode voltage is changed. This is indicative of a common-mode gain, something we're trying to avoid. More than that, its a common-mode gain of our own making, having nothing to do with imperfections in the op-amps themselves. With a much-tempered differential gain (actually equal to 3 in this particular circuit) and no negative feedback outside the circuit, this common-mode gain will go unchecked in an instrument signal application.

There is only one way to correct this common-mode gain, and that is to balance all the resistor values. When designing an instrumentation amplifier from discrete components (rather than purchasing one in an integrated package), it is wise to provide some means of making fine adjustments to at least one of the four resistors connected to the final op-amp to be able to "trim away" any such common-mode gain. Providing the means to "trim" the resistor network has additional benefits as well. Suppose that all resistor values are exactly as they should be, but a common-mode gain exists due to an imperfection in one of the op-amps. With the adjustment provision, the resistance could be trimmed to compensate for this unwanted gain.

One quirk of some op-amp models is that of output *latch-up*, usually caused by the common-mode input voltage exceeding allowable limits. If the common-mode voltage falls outside of the manufacturer's specified limits, the output may suddenly "latch" in the high mode (saturate at full output voltage). In JFET-input operational amplifiers, latch-up may occur if the common-mode input voltage approaches too closely to the negative power supply rail voltage. On the TL082 op-amp, for example, this occurs when the common-mode input voltage comes within about 0.7 volts of the negative power supply rail voltage. Such a situation may easily occur in a single-supply circuit, where the negative power supply rail is ground (0 volts), and the input signal is free to swing to 0 volts.

Latch-up may also be triggered by the common-mode input voltage *exceeding* power supply rail voltages, negative or positive. As a rule, you should never allow either input voltage to rise above the positive power supply rail voltage, or sink below the negative power supply rail voltage, even if the op-amp in question is protected against latch-up (as are the 741 and 1458 op-amp models). At the very least, the op-amp's behavior may become unpredictable. At worst, the kind of latch-up triggered by input voltages exceeding power supply voltages may be destructive to the op-amp.

While this problem may seem easy to avoid, its possibility is more likely than you might think. Consider the case of an operational amplifier circuit during power-up. If the circuit receives full input signal voltage *before* its own power supply has had time enough to charge the filter capacitors, the common-mode input voltage may easily exceed the power supply rail voltages for a short time.

If the op-amp receives signal voltage from a circuit supplied by a different power source, and its own power source fails, the signal voltage(s) may exceed the power supply rail voltages for an indefinite amount of time!

Offset Voltage

Another practical concern for op-amp performance is *voltage offset*. That is, effect of having the output voltage something other than zero volts when the two input terminals are shorted together. Remember that operational amplifiers are differential amplifiers above all: they're supposed to amplify the difference in voltage between the two input connections and nothing more.

When that input voltage difference is exactly zero volts, we would (ideally) expect to have exactly zero volts present on the output. However, in the real world this rarely happens. Even if the op-amp in question has zero common-mode gain (infinite CMRR), the output voltage may not be at zero when both inputs are shorted together. This deviation from zero is called *offset*.

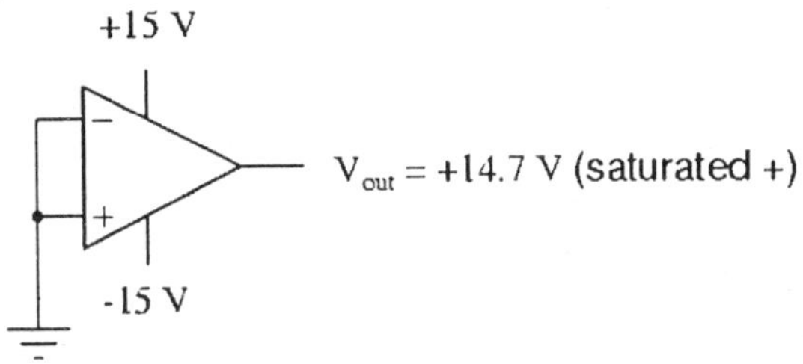

A perfect op-amp would output exactly zero volts with both its inputs shorted together and grounded. However, most op-amps off the shelf will drive their outputs to a saturated level, either negative or positive. In the example shown above, the output voltage is saturated at a value of positive 14.7 volts, just a bit less than +V (+15 volts) due to the positive saturation limit of this particular op-amp. Because the offset in this op-amp is driving the output to a completely saturated point, there's no way of telling how much voltage offset is present at the output. If the +V/-V split power supply was of a high enough voltage, who knows, maybe the output would be several hundred volts one way or the other due to the effects of offset.

For this reason, offset voltage is usually expressed in terms of the equivalent amount of *input* voltage differential producing this effect. In other words, we imagine that the op-amp is perfect (no offset whatsoever), and a small voltage is being applied in series with one of the inputs to force the output voltage one way or the other away from zero. Being that op-amp differential gains are so high, the figure for "input offset voltage" doesn't have to be much to account for what we see with shorted inputs:

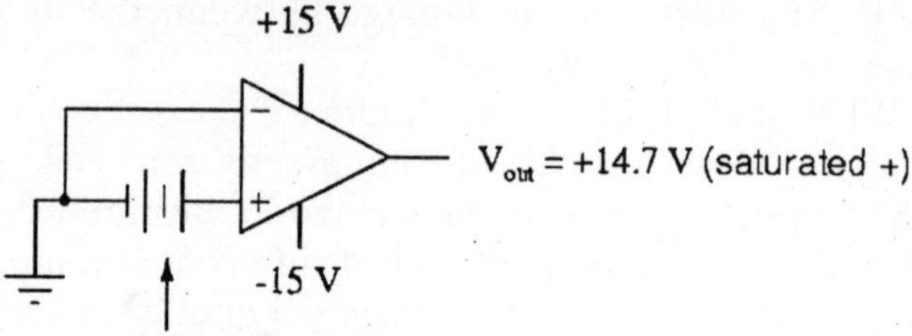

Input offset voltage
(internal to the real op-amp, external to this ideal op-amp)

Offset voltage will tend to introduce slight errors in any op-amp circuit. So how do we compensate for it? Unlike common-mode gain, there are usually provisions made by the manufacturer to trim the offset of a packaged op-amp. Usually, two extra terminals on the op-amp package are reserved for connecting an external "trim" potentiometer. These connection points are labeled *offset null* and are used in this general way:

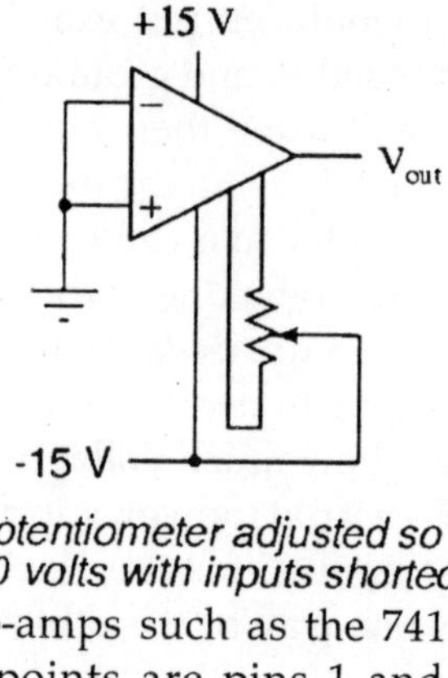

Potentiometer adjusted so that V_{out} = 0 volts with inputs shorted together

On single op-amps such as the 741 and 3130, the offset null connection points are pins 1 and 5 on the 8-pin DIP package. Other models of op-amp may have the offset null connections located on different pins, and/or require a slightly difference configuration of trim potentiometer connection. Some op-amps don't provide offset null pins at all! Consult the manufacturer's specifications for details.

Bias Current

Inputs on an op-amp have extremely high input impedances. That is, the input currents entering or exiting an op-amp's two input signal connections are extremely small. For most purposes of op-amp circuit analysis, we treat them as though they don't exist at all. We analyze the circuit as though there was absolutely zero current entering or exiting the input connections.

This idyllic picture, however, is not entirely true. Op-amps, especially those op-amps with bipolar transistor inputs, have to have some amount of current through their input connections in order for their internal circuits to be properly biased. These currents, logically, are called *bias currents*. Under certain conditions, op-amp bias currents may be problematic. The following circuit illustrates one of those problem conditions:

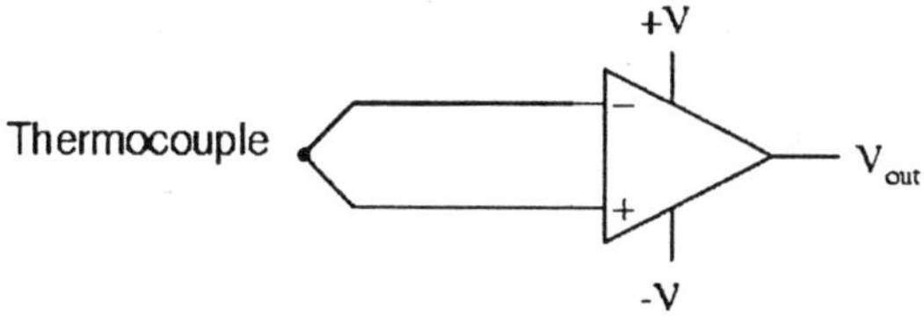

At first glance, we see no apparent problems with this circuit. A thermocouple, generating a small voltage proportional to temperature (actually, a voltage proportional to the *difference* in temperature between the measurement junction and the "reference" junction formed when the alloy thermocouple wires connect with the copper wires leading to the op-amp) drives the op-amp either positive or negative. In other words, this is a kind of comparator circuit, comparing the temperature between the end thermocouple junction and the reference junction (near the op-amp). The problem is this: the wire loop formed by the thermocouple does not provide a path for both input bias currents, because both bias currents are trying to go the same way (either into the op-amp or out of it).

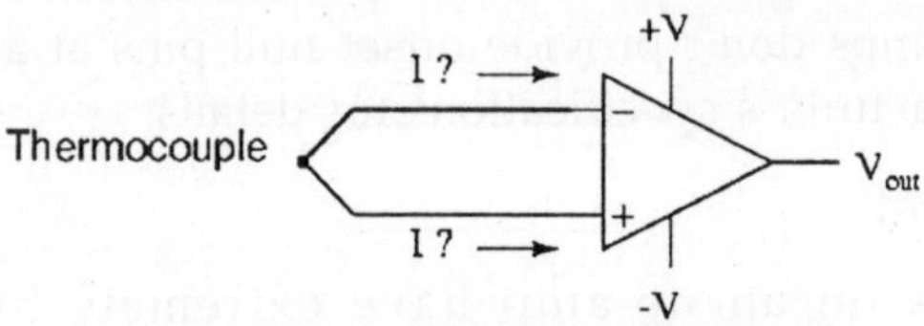

*This comparator circuit **won't** work*

In order for this circuit to work properly, we must ground one of the input wires, thus providing a path to (or from) ground for both currents:

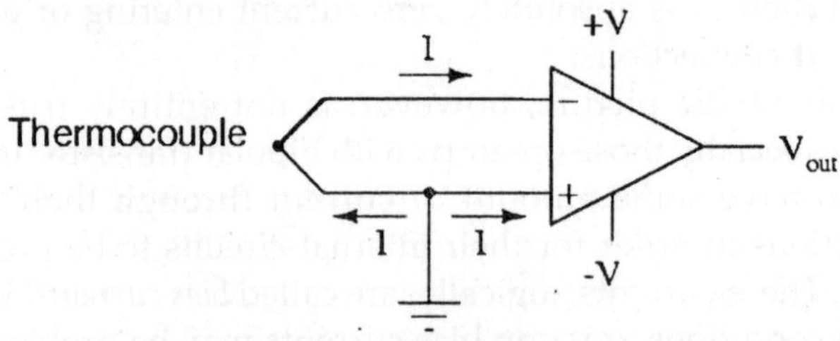

*This comparator circuit **will** work*

Not necessarily an obvious problem, but a very real one!

Another way input bias currents may cause trouble is by dropping unwanted voltages across circuit resistances. Take this circuit for example:

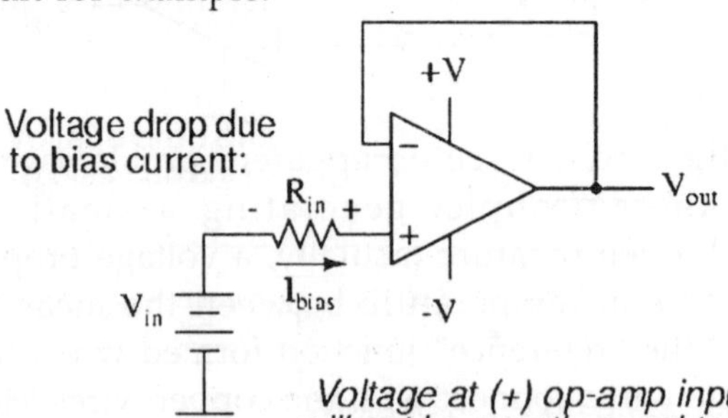

Voltage at (+) op-amp input will not be exactly equal to V_{in}

We expect a voltage follower circuit such as the one above to reproduce the input voltage precisely at the output. But what about the resistance in series with the input voltage source? If there is any bias current through the noninverting

(+) input at all, it will drop some voltage across R_{in}, thus making the voltage at the noninverting input unequal to the actual V_{in} value.

Bias currents are usually in the microamp range, so the voltage drop across R_{in} won't be very much, unless R_{in} is very large.

One example of an application where the input resistance (R_{in}) *would* be very large is that of pH probe electrodes, where one electrode contains an ion-permeable glass barrier (a very poor conductor, with millions of Ω of resistance).

If we were actually building an op-amp circuit for pH electrode voltage measurement, we'd probably want to use a FET or MOSFET (IGFET) input op-amp instead of one built with bipolar transistors (for less input bias current). But even then, what slight bias currents may remain can cause measurement errors to occur, so we have to find some way to mitigate them through good design.

One way to do so is based on the assumption that the two input bias currents will be the same. In reality, they are often close to being the same, the difference between them referred to as the *input offset current*. If they are the same, then we should be able to cancel out the effects of input resistance voltage drop by inserting an equal amount of resistance in series with the other input, like this:

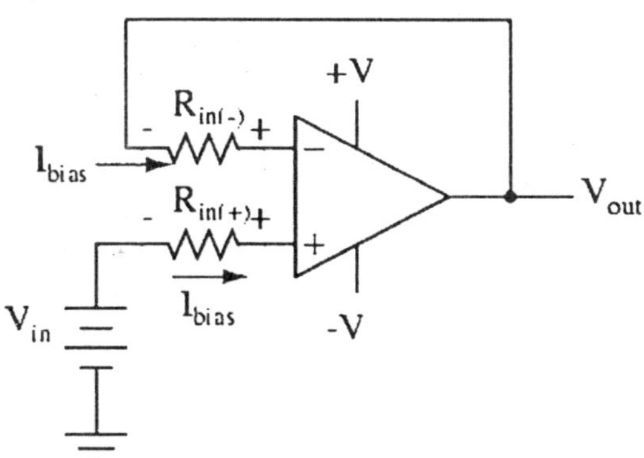

With the additional resistance added to the circuit, the output voltage will be closer to V_{in} than before, even if there is some offset between the two input currents.

For both inverting and noninverting amplifier circuits, the bias current compensating resistor is placed in series with the noninverting (+) input to compensate for bias current voltage drops in the divider network:

Noninverting amplifier with compensating resistor

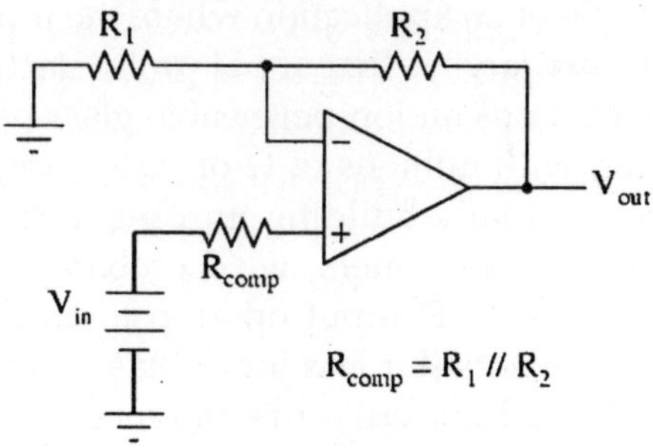

$R_{comp} = R_1 // R_2$

Inverting amplifier with compensating resistor

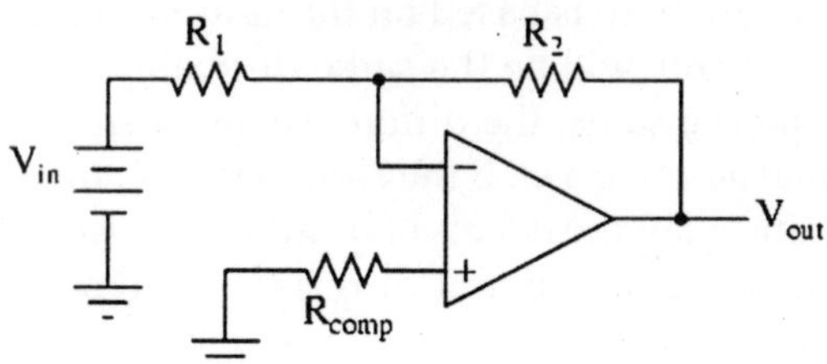

$R_{comp} = R_1 // R_2$

In either case, the compensating resistor value is determined by calculating the parallel resistance value of R_1 and R_2. Why is the value equal to the *parallel* equivalent of R_1 and R_2? When using the Superposition Theorem to figure how much voltage drop will be produced by the inverting (-) input's bias current, we treat the bias current as though it were coming from a current source inside the op-amp and short-circuit all voltage sources (V_{in} and V_{out}). This gives two parallel paths for bias current (through R_1 and through R_2, both to ground). We want to duplicate the bias current's effect on the noninverting (+) input, so the resistor value we choose to insert in series with that input needs to be equal to R_1 in parallel with R_2.

A related problem, occasionally experienced by students just learning to build operational amplifier circuits, is caused by a lack of a common ground connection to the power supply. It is *imperative* to proper op-amp function that some terminal of the DC power supply be common to the "ground" connection of the input signal(s). This provides a complete path for the bias currents, feedback current(s), and for the load (output) current. Take this circuit illustration, for instance, showing a properly grounded power supply:

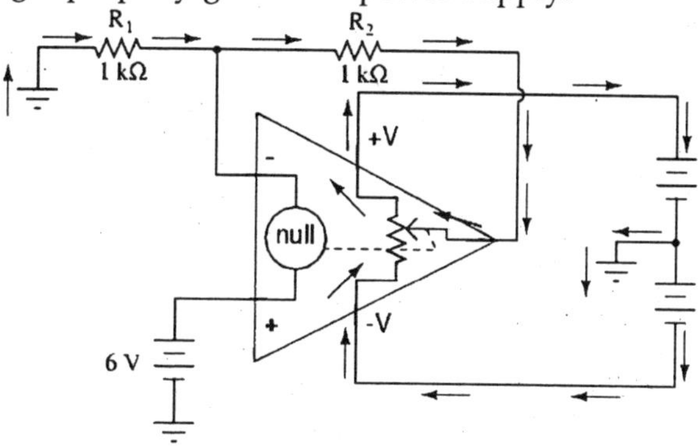

Here, arrows denote the path of electron flow through the power supply batteries, both for powering the op-amp's internal circuitry (the "potentiometer" inside of it that controls output voltage), and for powering the feedback loop of resistors R_1 and R_2. Suppose, however, that the ground connection for this "split" DC power supply were to be removed. The effect of doing this is profound:

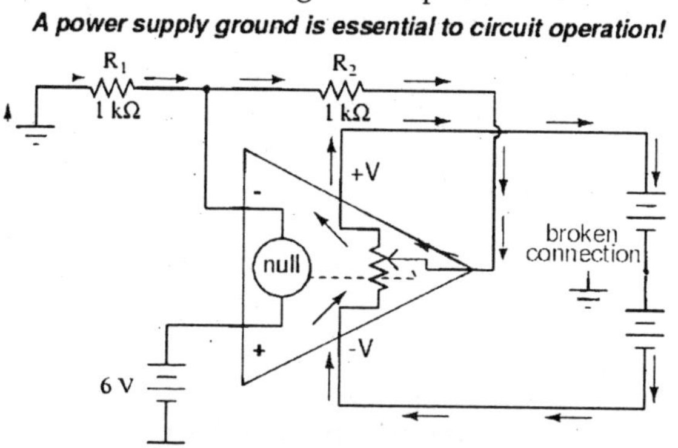

No electrons may flow in or out of the op-amp's output terminal, because the pathway to the power supply is a "dead end." Thus, no electrons flow through the ground connection to the left of R_1, neither through the feedback loop. This effectively renders the op-amp useless: it can neither sustain current through the feedback loop, nor through a grounded load, since there is no connection from any point of the power supply to ground.

The bias currents are also stopped, because they rely on a path to the power supply and back to the input source through ground. The following diagram shows the bias currents (only), as they go through the input terminals of the op-amp, through the base terminals of the input transistors, and eventually through the power supply terminal(s) and back to ground.

Bias current paths shown, through power supply

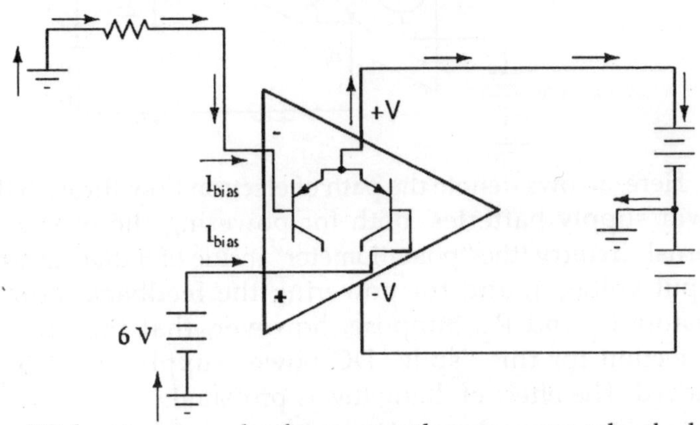

Without a ground reference on the power supply, the bias currents will have no complete path for a circuit, and they will halt. Since bipolar junction transistors are current-controlled devices, this renders the input stage of the op-amp useless as well, as both input transistors will be forced into cutoff by the complete lack of base current.

Drift

Being semiconductor devices, op-amps are subject to slight changes in behavior with changes in operating temperature. Any changes in op-amp performance with temperature fall under the category of op-amp *drift*. Drift parameters can be specified for bias currents, offset voltage, and the like. Consult the manufacturer's data sheet for specifics on any particular op-amp.

To minimize op-amp drift, we can select an op-amp made to have minimum drift, and/or we can do our best to keep the operating temperature as stable as possible. The latter action may involve providing some form of temperature control for the inside of the equipment housing the op-amp(s). This is not as strange as it may first seem. Laboratory-standard precision voltage reference generators, for example, are sometimes known to employ "ovens" for keeping their sensitive components (such as zener diodes) at constant temperatures. If extremely high accuracy is desired over the usual factors of cost and flexibility, this may be an option worth looking at.

Frequency Response

With their incredibly high differential voltage gains, op-amps are prime candidates for a phenomenon known as *feedback oscillation*. You've probably heard the equivalent audio effect when the volume (gain) on a public-address or other microphone amplifier system is turned too high: that high pitched squeal resulting from the sound waveform "feeding back" through the microphone to be amplified again. An op-amp circuit can manifest this same effect, with the feedback happening electrically rather than audibly.

A case example of this is seen in the 3130 op-amp, if it is connected as a voltage follower with the bare minimum of wiring connections (the two inputs, output, and the power supply connections). The output of this op-amp will self-oscillate due to its high gain, no matter what the input voltage. To combat this, a small *compensation capacitor* must be connected to two specially-provided terminals on the op-amp.

The capacitor provides a high-impedance path for negative feedback to occur within the op-amp's circuitry, thus decreasing the AC gain and inhibiting unwanted oscillations. If the op-amp is being used to amplify high-frequency signals, this compensation capacitor may not be needed, but it is absolutely essential for DC or low-frequency AC signal operation.

Some op-amps, such as the model 741, have a compensation capacitor built in to minimize the need for external components. This improved simplicity is not without a cost: due to that capacitor's presence inside the op-amp, the negative feedback tends to get stronger as the operating frequency increases (that capacitor's reactance decreases with higher frequencies). As a result, the op-amp's differential voltage gain decreases as frequency goes up: it becomes a less effective amplifier at higher frequencies.

Op-amp manufacturers will publish the frequency response curves for their products. Since a sufficiently high differential gain is absolutely essential to good feedback operation in op-amp circuits, the gain/frequency response of an op-amp effectively limits its "bandwidth" of operation. The circuit designer must take this into account if good performance is to be maintained over the required range of signal frequencies.

OPERATIONAL AMPLIFIER MODELS

While mention of operational amplifiers typically provokes visions of semiconductor devices built as integrated circuits on a miniature silicon chip, the first op-amps were actually vacuum tube circuits. The first commercial, general purpose operational amplifier was manufactured by the George

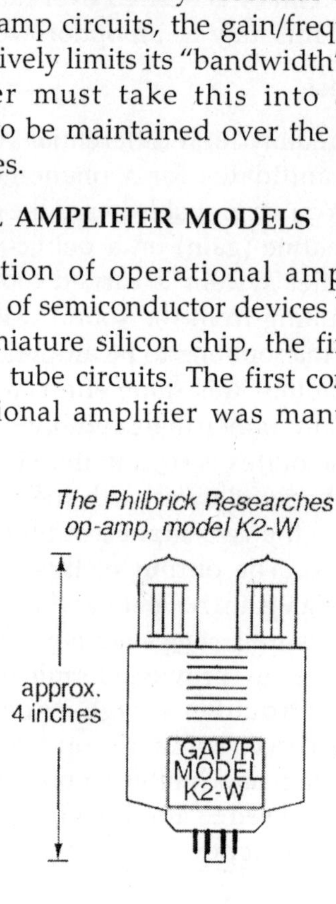

A. Philbrick Researches, Incorporated, in 1952. Designated the K2-W, it was built around two twin-triode tubes mounted in an assembly with an octal (8-pin) socket for easy installation and servicing in electronic equipment chassis of that era. The assembly looked something like this:

The schematic diagram shows the two tubes, along with ten resistors and two capacitors, a fairly simple circuit design even by 1952 standards:

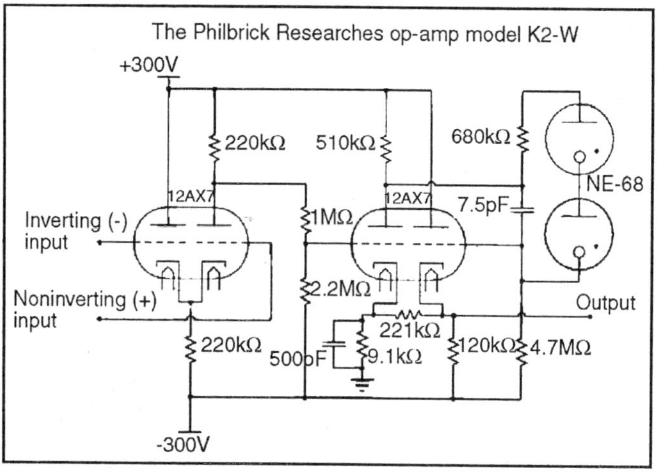

In case you're unfamiliar with the operation of vacuum tubes, they operate similarly to N-channel depletion-type IGFET transistors: that is, they conduct more current when the control grid (the dashed line) is made more positive with respect to the cathode (the bent line near the bottom of the tube symbol), and conduct less current when the control grid is made less positive (or more negative) than the cathode. The twin triode tube on the left functions as a *differential pair*, converting the differential inputs (inverting and noninverting input voltage signals) into a single, amplified voltage signal which is then fed to the control grid of the left triode of the second triode pair through a voltage divider (1 MΩ – 2.2 MΩ). That triode amplifies and inverts the output of the differential pair for a larger voltage gain, then the amplified signal is coupled to the second triode of the same dual-triode tube in a noninverting amplifier configuration for a larger current gain.

The two neon "glow tubes" act as voltage regulators, similar to the behavior of semiconductor zener diodes, to provide a bias voltage in the coupling between the two single-ended amplifier triodes.

With a dual-supply voltage of +300/-300 volts, this op-amp could only swing its output +/- 50 volts, which is very poor by today's standards. It had an open-loop voltage gain of 15,000 to 20,000, a slew rate of +/- 12 volts/µsecond, a maximum output current of 1 mA, a quiescent power consumption of over 3 watts (not including power for the tubes' filaments!), and cost about $24 in 1952 dollars. Better performance could have been attained using a more sophisticated circuit design, but only at the expense of greater power consumption, greater cost, and decreased reliability.

With the advent of solid-state transistors, op-amps with far less quiescent power consumption and increased reliability became feasible, but many of the other performance parameters remained about the same. Take for instance Philbrick's model P55A, a general-purpose solid-state op-amp circa 1966.

The P55A sported an open-loop gain of 40,000, a slew rate of 1.5 volt/µsecond and an output swing of +/- 11 volts (at a power supply voltage of +/- 15 volts), a maximum output current of 2.2 mA, and a cost of $49 (or about $21 for the "utility grade" version).

The P55A, as well as other op-amps in Philbrick's lineup of the time, was of discrete-component construction, its constituent transistors, resistors, and capacitors housed in a solid "brick" resembling a large integrated circuit package.

It isn't very difficult to build a crude operational amplifier using discrete components. A schematic of one such circuit is shown in Figure.

While its performance is rather dismal by modern standards, it demonstrates that complexity is not necessary to create a minimally functional op-amp. Transistors Q_3 and Q_4 form the heart of another differential pair circuit, the semiconductor equivalent of the first triode tube in the K2-W schematic.

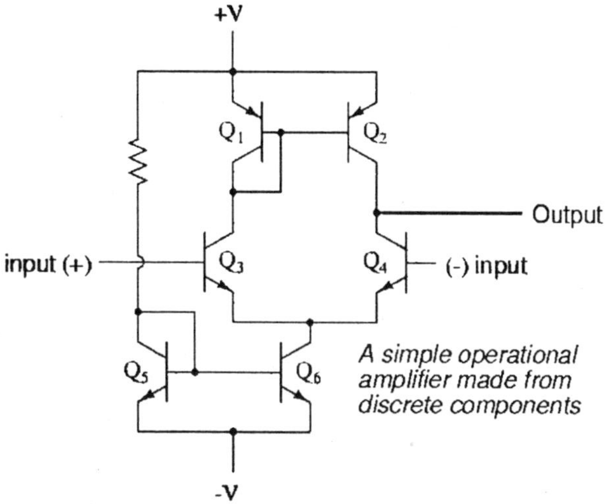

Fig A Simple Operational Amplifier Made from Discrete Components.

As it was in the vacuum tube circuit, the purpose of a differential pair is to amplify and convert a differential voltage between the two input terminals to a single-ended output voltage. With the advent of integrated-circuit (IC) technology, op-amp designs experienced a dramatic increase in performance, reliability, density, and economy. Between the years of 1964 and 1968.

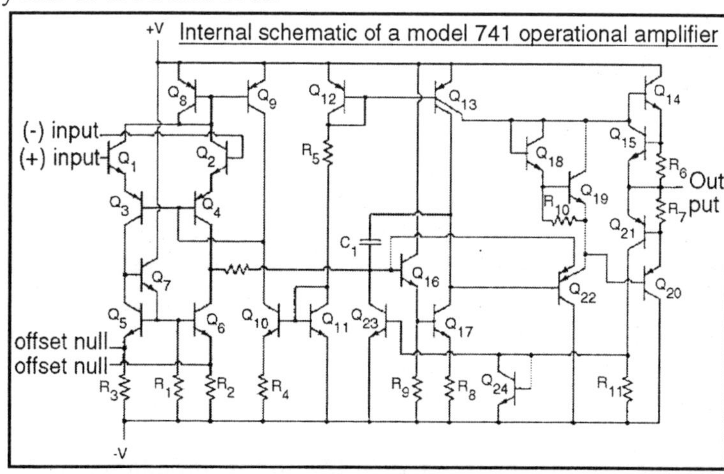

Fig. Schematic Diagram of a Model 741 op-amp.

The Fairchild corporation introduced three models of IC op-amps: the 702, 709, and the still-popular 741. While the 741 is now considered outdated in terms of performance, it is still a favorite among hobbyists for its simplicity and fault tolerance (short-circuit protection on the output, for instance). Personal experience abusing many 741 op-amps has led me to the conclusion that it is a hard chip to kill.

The internal schematic diagram for a model 741 op-amp is shown in Figure above

By integrated circuit standards, the 741 is a very simple device: an example of *small-scale integration,* or *SSI* technology. It would be no small matter to build this circuit using discrete components, so you can see the advantages of even the most primitive integrated circuit technology over discrete components where high parts counts are involved.

For the hobbyist, student, or engineer desiring greater performance, there are literally hundreds of op-amp models to choose from. Many sell for less than a dollar apiece, even retail! Special-purpose instrumentation and radio-frequency (RF) op-amps may be quite a bit more expensive. In this section I will showcase several popular and affordable op-amps, comparing and contrasting their performance specifications. The venerable 741 is included as a "benchmark" for comparison, although it is, as I said before, considered an obsolete design.

Listed in Table above are but a few of the low-cost operational amplifier models widely available from electronics suppliers. Most of them are available through retail supply stores such as Radio Shack. All are under $1.00 cost direct from the manufacturer (year 2001 prices). As you can see, there is substantial variation in performance between some of these units. Take for instance the parameter of input bias current: the CA3130 wins the prize for lowest, at 0.05 nA (or 50 pA), and the LM833 has the highest at slightly over 1 μA.

The model CA3130 achieves its incredibly low bias current through the use of MOSFET transistors in its input stage. One manufacturer advertises the 3130's input impedance as 1.5 tera-ohms, or $1.5 \times 10^{12}\ \Omega$! Other op-amps shown here with low bias current figures use JFET input transistors, while the high bias current models use bipolar input transistors.

Table: Widely Used Operational Amplifiers

number	(count)	(V)	(Mhz)	(nA)	(V/μS)	(mA)
Model	Devices/	Power	Bandwidth	Bias	Slew	Output
	package	supply		current	rate	current
number	(count)	(V)	(Mhz)	(nA)	(V/μS)	(mA)
TL082	2	12/36	4	8	13	17
LM301A	1	10/36	1	250	0.5	25
LM318	1	10/40	15	500	70	20
LM324	4	3/32	1	45	0.25	20
LF353	2	12/36	4	8	13	20
LF356	1	10/36	5	8	12	25
LF411	1	10/36	4	20	15	25
741C	1	10/36	1	500	0.5	25
LM833	2	10/36	15	1050	7	40
LM1458	2	6/36	1	800	10	45
CA3130	1	5/16	15	0.05	10	20

If low power supply voltage is a requirement, I recommend the model 324, as it functions on as low as 3 volts DC. Its input bias current requirements are also low, and it provides four op-amps in a single 14-pin chip. Its major weakness is speed, limited to 1 MHz bandwidth and an output slew rate of only 0.25 volts per μs. For high-frequency AC amplifier circuits, the 318 is a very good "general purpose" model.

Special-purpose op-amps are available for modest cost which provide better performance specifications. Many of these are tailored for a specific type of performance advantage, such as maximum bandwidth or minimum bias current. Take for instance the op-amps, both designed for high bandwidth in Table below.

Table: High Bandwidth Operational Amplifiers

Model number	Devices/ package (count)	Power supply (V)	Bandwidth (Mhz)	Bias current (nA)	Slew rate (V/μS)	Output current (mA)
CLC404	1	10/14	232	44,000	2600	70
CLC425	1	5/14	1900	40,000	350	90

The CLC404 lists at $21.80 (almost as much as George Philbrick's first commercial op-amp, albeit without correction for inflation), while the CLC425 is quite a bit less expensive at $3.23 per unit. In both cases high speed is achieved at the expense of high bias currents and restrictive power supply voltage ranges. Some op-amps, designed for high power output are listed in Table below.

Table: High Current Operational Amplifiers

Model number	Devices/ package (count)	Power supply (V)	Bandwidth (Mhz)	Bias current (nA)	Slew rate (V/μS)	Output current (mA)
LM12CL	1	15/80	0.7	1000	9	13,000
LM7171	1	5.5/36	200	12,000	4100	100

Yes, the LM12CL actually has an output current rating of *13 amps* (13,000 milliamps)! It lists at $14.40, which is not a lot of money, considering the raw power of the device. The LM7171, on the other hand, trades high current output ability for fast voltage output ability (a high slew rate). It lists at $1.19, about as low as some "general purpose" op-amps.

Amplifier packages may also be purchased as complete application circuits as opposed to bare operational amplifiers. The Burr-Brown and Analog Devices corporations, for example, both long known for their precision amplifier product lines, offer instrumentation amplifiers in pre-designed packages as well as other specialized amplifier devices. In designs where high precision and repeatability after repair is important, it might be advantageous for the circuit designer to choose such a pre-engineered amplifier "block" rather than build the circuit from individual op-amps. Of course, these units typically cost quite a bit more than individual op-amps.

Chapter 7

Electron Tubes

An often neglected area of study in modern electronics is that of *tubes*, more precisely known as *vacuum tubes* or *electron tubes*. Almost completely overshadowed by semiconductor, or "solid-state" components in most modern applications, tube technology once dominated electronic circuit design.

In fact, the historical transition from "electric" to "electronic" circuits really began with tubes, for it was with tubes that we entered into a whole new realm of circuit function:

A way of controlling the flow of electrons (current) in a circuit by means of another electric signal (in the case of most tubes, the controlling signal is a small voltage). The semiconductor counterpart to the tube, of course, is the transistor.

Transistors perform much the same function as tubes: controlling the flow of electrons in a circuit by means of another flow of electrons in the case of the bipolar transistor, and controlling the flow of electrons by means of a voltage in the case of the field-effect transistor. In either case, a relatively small electric signal controls a relatively large electric current. This is the essence of the word "electronic," so as to distinguish it from "electric," which has more to do with how electron flow is regulated by Ohm's Law and the physical attributes of wire and components.

Though tubes are now obsolete for all but a few specialized applications, they are still a worthy area of study. If nothing else, it is fascinating to explore "the way things used to be done" in order to better appreciate modern technology.

EARLY TUBE HISTORY

Thomas Edison, that prolific American inventor, is often credited with the invention of the incandescent lamp. More accurately, it could be said that Edison was the man who *perfected* the incandescent lamp. Edison's successful design of 1879 was actually preceded by 77 years by the British scientist Sir Humphry Davy, who first demonstrated the principle of using electric current to heat a thin strip of metal (called a "filament") to the point of incandescence (glowing white hot).

Edison was able to achieve his success by placing his filament (made of carbonized sewing thread) inside of a clear glass bulb from which the air had been forcibly removed. In this vacuum, the filament could glow at white-hot temperatures without being consumed by combustion:

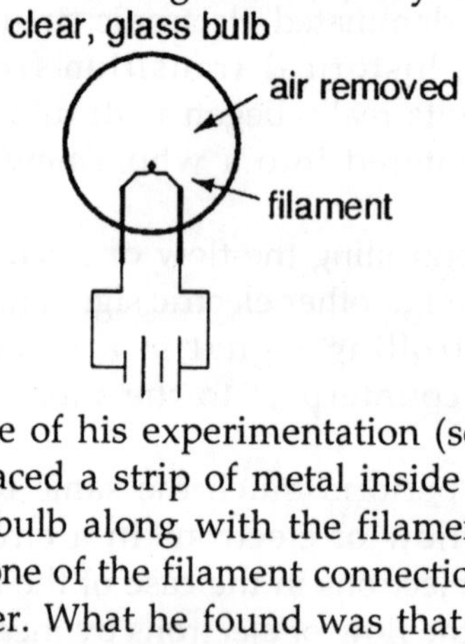

In the course of his experimentation (sometime around 1883), Edison placed a strip of metal inside of an evacuated (vacuum) glass bulb along with the filament. Between this metal strip and one of the filament connections he attached a sensitive ammeter. What he found was that electrons would flow through the meter whenever the filament was hot, but ceased when the filament cooled down:

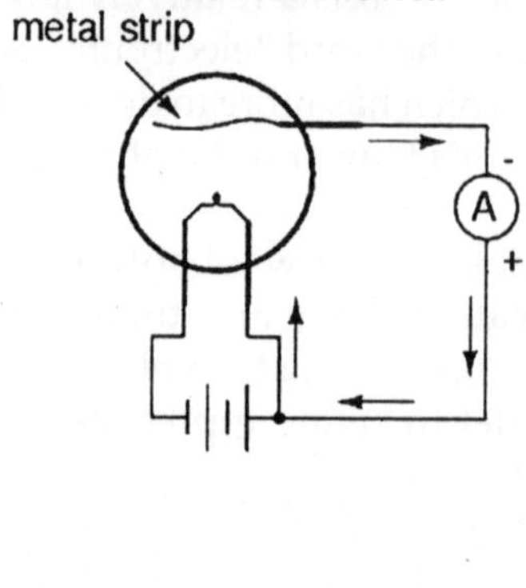

The white-hot filament in Edison's lamp was liberating free electrons into the vacuum of the lamp, those electrons finding their way to the metal strip, through the galvanometer, and back to the filament. His curiosity piqued, Edison then connected a fairly high-voltage battery in the galvanometer circuit to aid the small current:

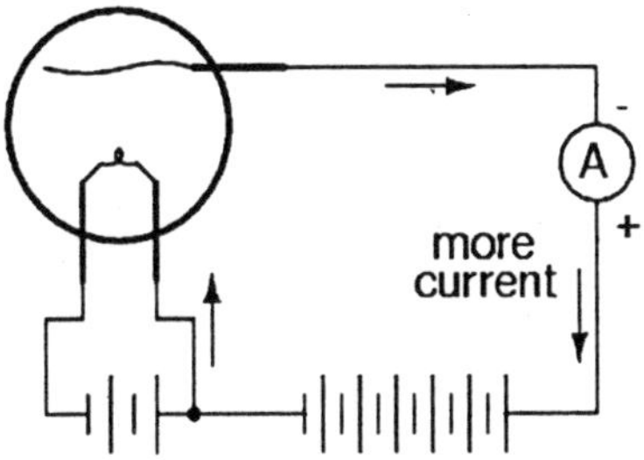

Sure enough, the presence of the battery created a much larger current from the filament to the metal strip. However, when the battery was turned around, there was little to no current at all!

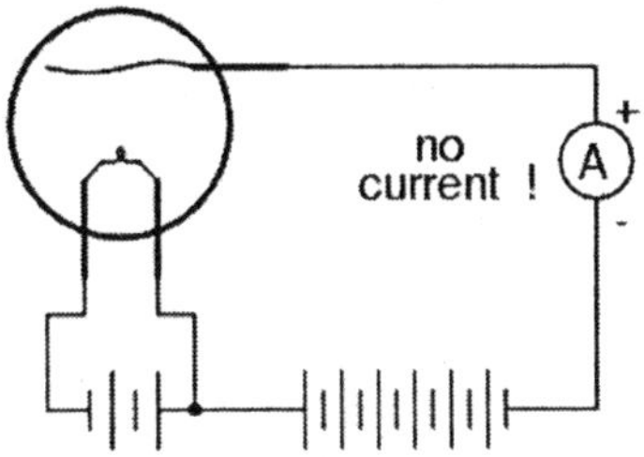

In effect, what Edison had stumbled upon was a diode! Unfortunately, he saw no practical use for such a device and proceeded with further refinements in his lamp design.

The one-way electron flow of this device (known as the *Edison Effect*) remained a curiosity until J. A. Fleming experimented with its use in 1895. Fleming marketed his device as a "valve," initiating a whole new area of study in electric circuits. Vacuum tube diodes — Fleming's "valves" being no exception — are not able to handle large amounts of current, and so Fleming's invention was impractical for any application in AC power, only for small electric signals.

Then in 1906, another inventor by the name of Lee De Forest started playing around with the "Edison Effect," seeing what more could be gained from the phenomenon. In doing so, he made a startling discovery: by placing a metal screen between the glowing filament and the metal strip (which by now had taken the form of a plate for greater surface area), the stream of electrons flowing from filament to plate could be regulated by the application of a small voltage between the metal screen and the filament:

The DeForest "Audion" tube

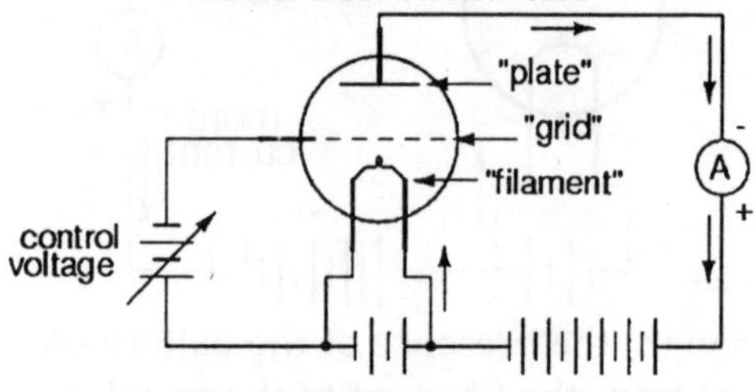

plate current can be controlled by the application of a small control voltage between the grid and filament!

De Forest called this metal screen between filament and plate a *grid*. It wasn't just the amount of voltage between grid and filament that controlled current from filament to plate, it was the polarity as well.

A negative voltage applied to the grid with respect to the filament would tend to choke off the natural flow of electrons, whereas a positive voltage would tend to enhance the flow. Although there was some amount of current through the grid, it was very small; much smaller than the current through the plate.

Perhaps most importantly was his discovery that the small amounts of grid voltage and grid current were having large effects on the amount of plate voltage (with respect to the filament) and plate current. In adding the grid to Fleming's "valve," De Forest had made the valve adjustable: it now functioned as an *amplifying* device, whereby a small electrical signal could take control over a larger electrical quantity.

The closest semiconductor equivalent to the Audion tube, and to all of its more modern tube equivalents, is an n-channel D-type MOSFET. It is a voltage-controlled device with a large current gain.

THE TRIODE

De Forest's Audion tube came to be known as the *triode* tube, because it had three elements: filament, grid, and plate. Later developments in diode tube technology led to the refinement of the electron emitter: instead of using the filament directly as the emissive element, another metal strip called the *cathode* could be heated by the filament.

This refinement was necessary in order to avoid some undesired effects of an incandescent filament as an electron emitter. First, a filament experiences a voltage drop along its length, as current overcomes the resistance of the filament material and dissipates heat energy. This meant that the voltage potential between different points along the length of the filament wire and other elements in the tube would not be constant.

For this and similar reasons, alternating current used as a power source for heating the filament wire would tend to introduce unwanted AC "noise" in the rest of the tube circuit. Furthermore, the surface area of a thin filament was limited at best, and limited surface area on the electron emitting element tends to place a corresponding limit on the tube's current-carrying capacity.

The cathode was a thin metal cylinder fitting snugly over the twisted wire of the filament. The cathode cylinder would be heated by the filament wire enough to freely emit electrons, without the undesirable side effects of actually carrying the heating current as the filament wire had to. The tube symbol for a triode with an indirectly-heated cathode looks like this:

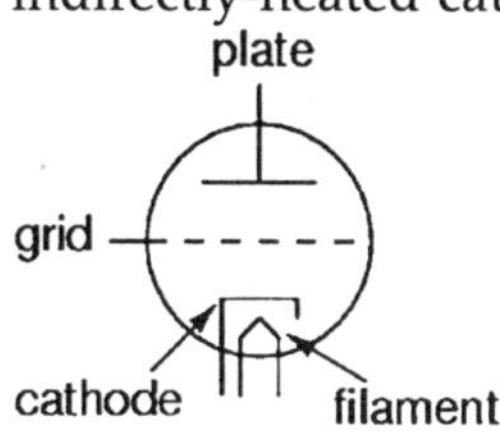

Since the filament is necessary for all but a few types of vacuum tubes, it is often omitted in the symbol for simplicity, or it may be included in the drawing but with no power connections drawn to it:

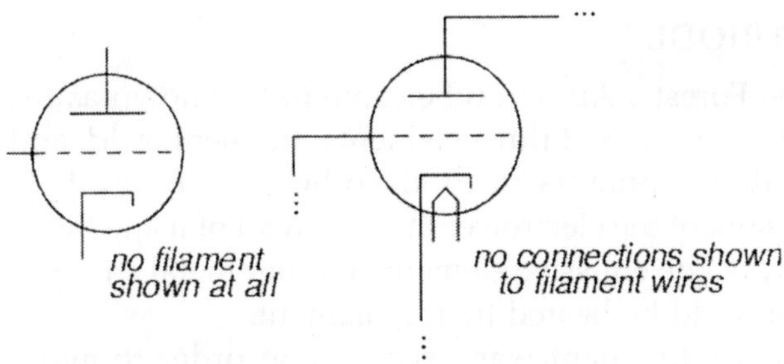

A simple triode circuit is shown to illustrate its basic operation as an amplifier:

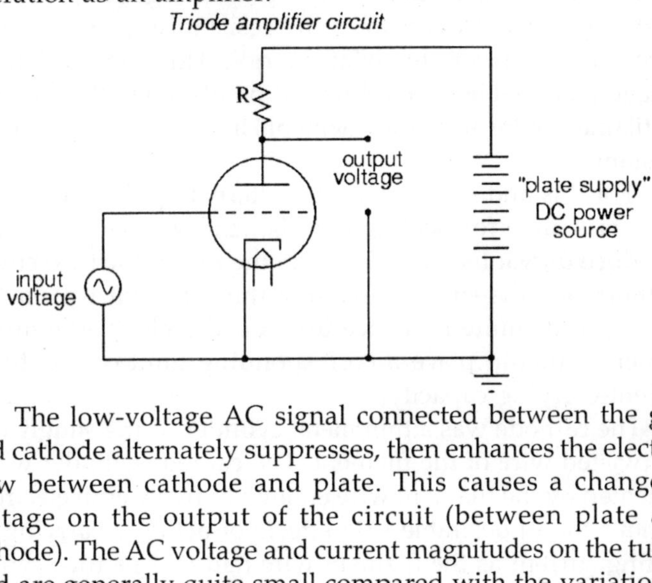

The low-voltage AC signal connected between the grid and cathode alternately suppresses, then enhances the electron flow between cathode and plate. This causes a change in voltage on the output of the circuit (between plate and cathode). The AC voltage and current magnitudes on the tube's grid are generally quite small compared with the variation of voltage and current in the plate circuit. Thus, the triode functions as an amplifier of the incoming AC signal (taking high-voltage, high-current DC power supplied from the large DC source on the right and "throttling" it by means of the tube's controlled conductivity).

In the triode, the amount of current from cathode to plate (the "controlled" current is a function both of grid-to-cathode voltage (the controlling signal) and the plate-to-cathode voltage (the electromotive force available to push electrons through the vacuum). Unfortunately, neither of these independent variables have a purely linear effect on the amount of current through the device (often referred to simply as the "plate current"). That is, triode current does not necessarily respond in a direct, proportional manner to the voltages applied.

In this particular amplifier circuit the nonlinearities are compounded, as plate voltage (with respect to cathode) changes along with the grid voltage (also with respect to cathode) as plate current is throttled by the tube. The result will be an output voltage waveform that doesn't precisely resemble the waveform of the input voltage. In other words, the quirkiness of the triode tube and the dynamics of this particular circuit will *distort* the waveshape. If we really wanted to get complex about how we stated this, we could say that the tube introduces *harmonics* by failing to exactly reproduce the input waveform.

Another problem with triode behavior is that of stray capacitance. Remember that any time we have two conductive surfaces separated by an insulating medium, a capacitor will be formed. Any voltage between those two conductive surfaces will generate an electric field within that insulating region, potentially storing energy and introducing reactance into a circuit. Such is the case with the triode, most problematically between the grid and the plate. It is as if there were tiny capacitors connected between the pairs of elements in the tube:

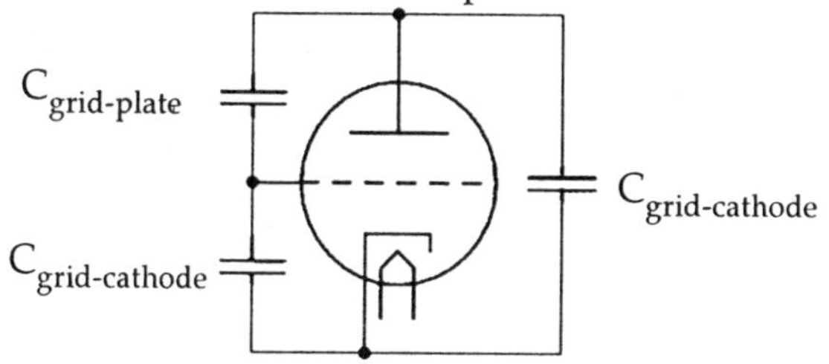

Now, this stray capacitance is quite small, and the reactive impedances usually high. Usually, that is, unless radio frequencies are being dealt with. As we saw with De Forest's Audion tube, radio was probably the prime application for this new technology, so these "tiny" capacitances became more than just a potential problem. Another refinement in tube technology was necessary to overcome the limitations of the triode.

THE TETRODE

As the name suggests, the *tetrode* tube contained four elements: cathode (with the implicit filament, or "heater"), grid, plate, and a new element called the *screen*. Similar in construction to the grid, the screen was a wire mesh or coil positioned between the grid and plate, connected to a source of positive DC potential (with respect to the cathode, as usual) equal to a fraction of the plate voltage.

When connected to ground through an external capacitor, the screen had the effect of electrostatically shielding the grid from the plate. Without the screen, the capacitive linking between the plate and the grid could cause significant signal feedback at high frequencies, resulting in unwanted oscillations.

The screen, being of less surface area and lower positive potential than the plate, didn't attract many of the electrons passing through the grid from the cathode, so the vast majority of electrons in the tube still flew by the screen to be collected by the plate:

Tetrode amplifier circuit

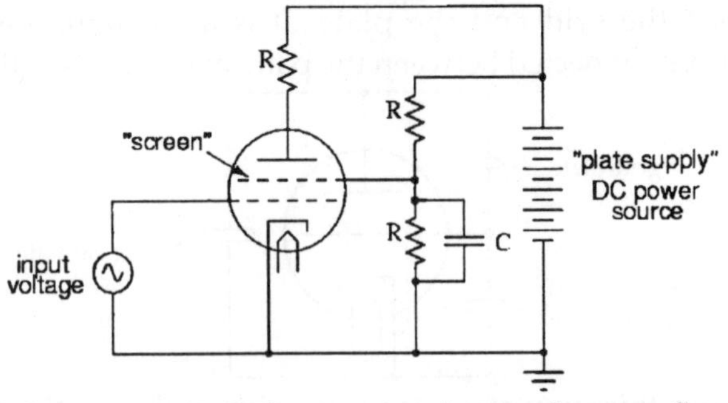

With a constant DC screen voltage, electron flow from cathode to plate became almost exclusively dependent upon grid voltage, meaning the plate voltage could vary over a wide range with little effect on plate current. This made for more stable gains in amplifier circuits, and better linearity for more accurate reproduction of the input signal waveform.

Despite the advantages realized by the addition of a screen, there were some disadvantages as well. The most significant disadvantage was related to something known as *secondary emission.*

When electrons from the cathode strike the plate at high velocity, they can cause free electrons to be jarred loose from atoms in the metal of the plate. These electrons, knocked off the plate by the impact of the cathode electrons, are said to be "secondarily emitted." In a triode tube, secondary emission is not that great a problem, but in a tetrode with a positively-charged screen grid in close proximity, these secondary electrons will be attracted to the screen rather than the plate from which they came, resulting in a loss of plate current. Less plate current means less gain for the amplifier, which is not good.

Two different strategies were developed to address this problem of the tetrode tube: *beam power* tubes and *pentodes.* Both solutions resulted in new tube designs with approximately the same electrical characteristics.

BEAM POWER TUBES

In the beam power tube, the basic four-element structure of the tetrode was maintained, but the grid and screen wires were carefully arranged along with a pair of auxiliary plates to create an interesting effect: focused beams or "sheets" of electrons traveling from cathode to plate.

These electron beams formed a stationary "cloud" of electrons between the screen and plate (called a "space charge") which acted to repel secondary electrons emitted from the plate back to the plate. A set of "beam-forming" plates, each connected to the cathode, were added to help maintain proper electron beam focus. Grid and screen wire coils were arranged in such a way that each turn or wrap of the screen fell directly behind a wrap of the grid, which placed the screen wires in the "shadow" formed by the grid. This precise alignment enabled the screen to still perform its shielding function with minimal interference to the passage of electrons from cathode to plate.

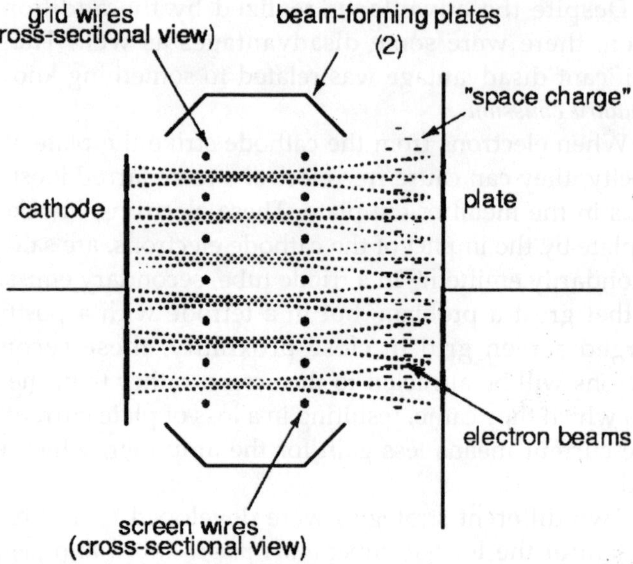

This resulted in lower screen current (and more plate current!) than an ordinary tetrode tube, with little added expense to the construction of the tube.

Beam power tetrodes were often distinguished from their non-beam counterparts by a different schematic symbol, showing the beam-forming plates:

The pentode tube

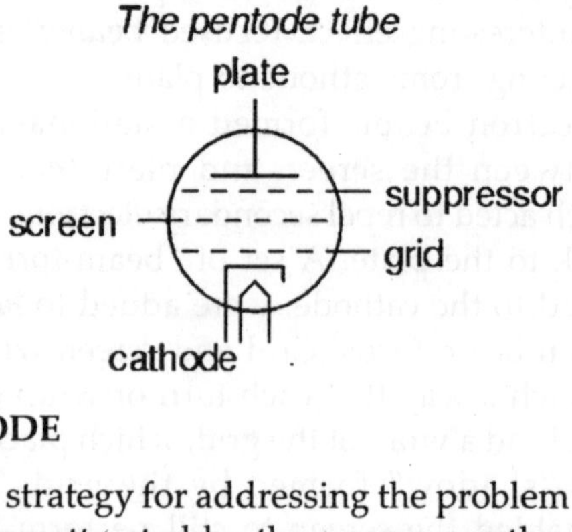

THE PENTODE

Another strategy for addressing the problem of secondary electrons being attracted by the screen was the addition of a fifth wire element to the tube structure: a *suppressor*. These five-element tubes were naturally called *pentodes*.

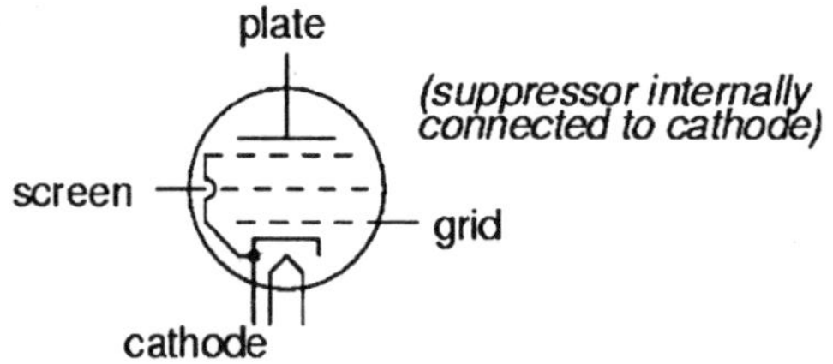

The suppressor was another wire coil or mesh situated between the screen and the plate, usually connected directly to ground potential. In some pentode tube designs, the suppressor was internally connected to the cathode so as to minimize the number of connection pins having to penetrate the tube envelope:

The suppressor's job was to repel any secondarily emitted electrons back to the plate: a structural equivalent of the beam power tube's space charge. This, of course, increased plate current and decreased screen current, resulting in better gain and overall performance. In some instances it allowed for greater operating plate voltage as well.

COMBINATION TUBES

Similar in thought to the idea of the integrated circuit, tube designers tried integrating different tube functions into single tube envelopes to reduce space requirements in more modern tube-type electronic equipment. A common combination seen within a single glass shell was two either diodes or two triodes. The idea of fitting pairs of diodes inside a single envelope makes a lot of sense in light of power supply full-wave rectifier designs, always requiring multiple diodes.

Of course, it would have been quite impossible to combine thousands of tube elements into a single tube envelope the way that thousands of transistors can be etched onto a single piece of silicon, but engineers still did their best to push the limits of tube miniaturization and consolidation. Some of these tubes, whimsically called *compactrons*, held four or more complete tube elements within a single envelope.

Sometimes the functions of two different tubes could be integrated into a single, combination tube in a way that simply worked more elegantly than two tubes ever could. An example of this was the *pentagrid converter,* more generally called a *heptode,* used in some superheterodyne radio designs. These tubes contained seven elements: 5 grids plus a cathode and a plate. Two of the grids were normally reserved for signal input, the other three relegated to screening and suppression (performance-enhancing) functions.

Combining the superheterodyne functions of oscillator and signal mixer together in one tube, the signal coupling between these two stages was intrinsic. Rather than having separate oscillator and mixer circuits, the oscillator creating an AC voltage and the mixer "mixing" that voltage with another signal, the pentagrid converter's oscillator section created an electron stream that oscillated in intensity which then directly passed through another grid for "mixing" with another signal.

This same tube was sometimes used in a different way: by applying a DC voltage to one of the control grids, the gain of the tube could be changed for a signal impressed on the other control grid. This was known as *variable-mu* operation, because the "mu" (μ) of the tube (its amplification factor, measured as a ratio of plate-to-cathode voltage change over grid-to-cathode voltage change with a constant plate current) could be altered at will by a DC control voltage signal.

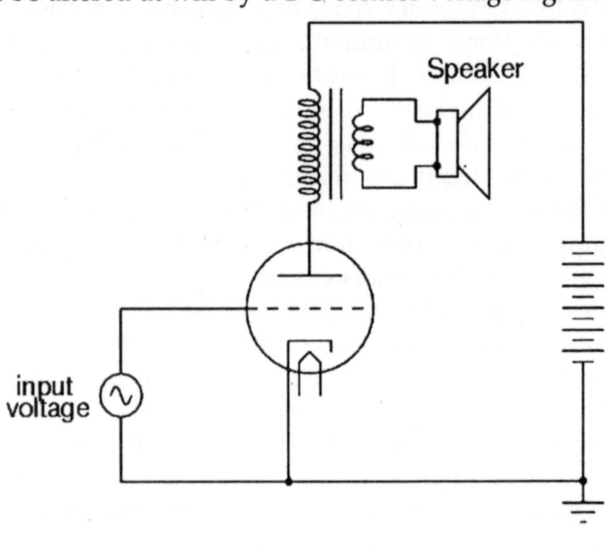

Enterprising electronics engineers also discovered ways to exploit such multi-variable capabilities of "lesser" tubes such as tetrodes and pentodes. One such way was the so-called *ultralinear* audio power amplifier, invented by a pair of engineers named.

Hafler and Keroes, utilizing a tetrode tube in combination with a "tapped" output transformer to provide substantial improvements in amplifier linearity (decreases in distortion levels). Consider a "single-ended" triode tube amplifier with an output transformer coupling power to the speaker:

If we substitute a tetrode for a triode in this circuit, we will see improvements in circuit gain resulting from the electrostatic shielding offered by the screen, preventing unwanted feedback between the plate and control grid:

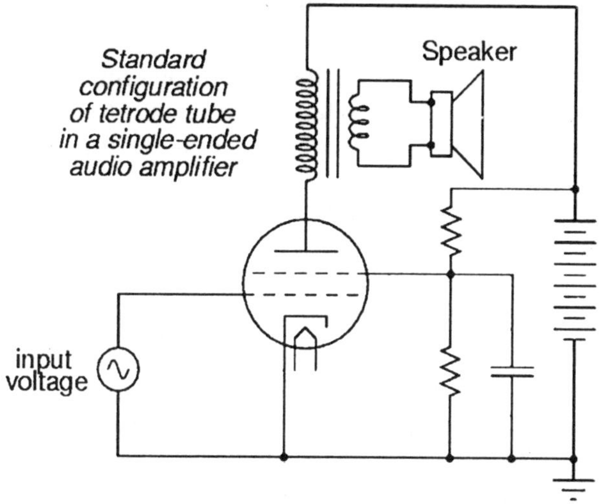

However, the tetrode's screen may be used for functions other than merely shielding the grid from the plate. It can also be used as another control element, like the grid itself. If a "tap" is made on the transformer's primary winding, and this tap connected to the screen, the screen will receive a voltage that varies with the signal being amplified (feedback). More specifically, the feedback signal is proportional to the rate-of-change of magnetic flux in the transformer core (dÖ/dt), thus improving the amplifier's ability to reproduce the input signal waveform at the speaker terminals and not just in the primary winding of the transformer:

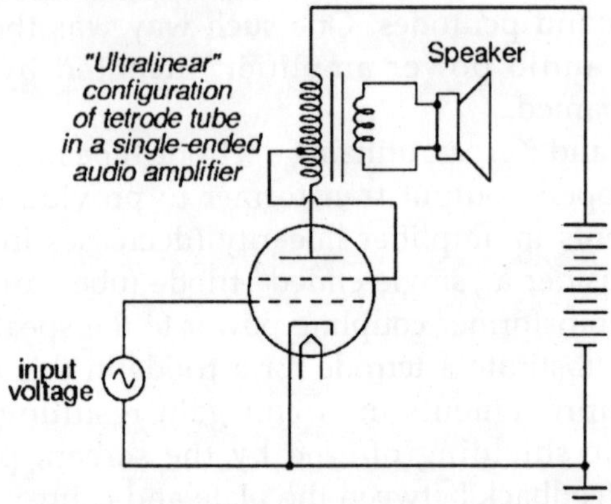

This signal feedback results in significant improvements in amplifier linearity (and consequently, distortion), so long as precautions are taken against "overpowering" the screen with too great a positive voltage with respect to the cathode. As a concept, the ultralinear (screen-feedback) design demonstrates the flexibility of operation granted by multiple grid-elements inside a single tube: a capability rarely matched by semiconductor components.

Some tube designs combined multiple tube functions in a most economic way: dual plates with a single cathode, the currents for each of the plates controlled by separate sets of control grids. Common examples of these tubes were *triode-heptode* and *triode-hexode* tubes (a hexode tube is a tube with four grids, one cathode, and one plate).

Other tube designs simply incorporated separate tube structures inside a single glass envelope for greater economy.

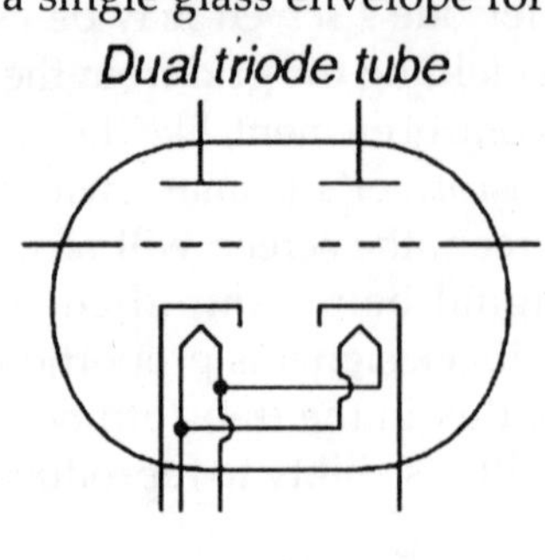

Dual diode (rectifier) tubes were quite common, as were dual triode tubes, especially when the power dissipation of each tube was relatively low.

The 12AX7 and 12AU7 models are common examples of dual-triode tubes, both of low-power rating. The 12AX7 is especially common as a preamplifier tube in electric guitar amplifier circuits.

TUBE PARAMETERS

For bipolar junction transistors, the fundamental measure of amplification is the Beta ratio (β), defined as the ratio of collector current to base current (I_C/I_B). Other transistor characteristics such as junction resistance, which in some amplifier circuits may impact performance as much as β, are quantified for the benefit of circuit analysis. Electron tubes are no different, their performance characteristics having been explored and quantified long ago by electrical engineers.

Before we can speak meaningfully on these characteristics, we must define several mathematical variables used for expressing common voltage, current, and resistance measurements as well as some of the more complex quantities:

μ = amplification factor, pronounced *"mu"* (unitless)
g_m = mutual conductance, in siemens
E_p = plate-to-cathode voltage
E_g = grid-to-cathode voltage
I_p = plate current
I_k = cathode current
E_s = input signal voltage
r_p = dynamic plate resistance, in ohms
D = delta, the Greek symbol for change

The two most basic measures of an amplifying tube's characteristics are its amplification factor (μ) and its mutual conductance (g_m), also known as *transconductance*. Transconductance is defined here just the same as it is for field-effect transistors, another category of voltage-controlled devices. Here are the two equations defining each of these performance characteristics:

$$m = \frac{\Delta E_p}{\Delta E_g} \text{ with constant } I_p \text{ (plate current)}$$

$$g_m = \frac{\Delta I_p}{\Delta E_g} \text{ with constant } E_p \text{ (plate voltage)}$$

Another important, though more abstract, measure of tube performance is its *plate resistance*. This is the measurement of plate voltage change over plate current change for a constant value of grid voltage. In other words, this is an expression of how much the tube acts like a resistor for any given amount of grid voltage, analogous to the operation of a JFET in its ohmic mode:

$$r_p = \frac{\Delta E_p}{\Delta I_p} \text{ with constant } E_g \text{ (grid voltate)}$$

The astute reader will notice that plate resistance may be determined by dividing the amplification factor by the transconductance:

$$\mu = \frac{\Delta E_p}{\Delta E_g} \qquad g_m = \frac{\Delta I_p}{\Delta E_g}$$

. . .dividing m by gm . . .

$$r_p = \frac{\dfrac{\Delta E_p}{\Delta E_g}}{\dfrac{\Delta I_p}{\Delta E_g}}$$

$$r_p = \frac{\Delta E_p}{\Delta E_g} \frac{\Delta E_g}{\Delta I_p}$$

$$r_p = \frac{\Delta E_g}{\Delta I_p}.$$

These three performance measures of tubes are subject to change from tube to tube (just as β ratios between two "identical" bipolar transistors are never precisely the same) and between different operating conditions. This variability is due partly to the unavoidable nonlinearities of electron tubes and partly due to how they are defined. Even supposing the existence of a perfectly linear tube, it will be impossible for all three of these measures to be constant over the allowable ranges of operation. Consider a tube that *perfectly* regulates current at any given amount of grid voltage (like a bipolar transistor with an absolutely constant β): that tube's plate resistance *must* vary with plate voltage, because plate current will not change even though plate voltage does.

Nevertheless, tubes were (and are) rated by these values at given operating conditions, and may have their characteristic curves published just like transistors.

IONIZATION (GAS-FILLED) TUBES

So far, we've explored tubes which are totally "evacuated" of all gas and vapor inside their glass envelopes, properly known as *vacuum tubes*. With the addition of certain gases or vapors, however, tubes take on significantly different characteristics, and are able to fulfill certain special roles in electronic circuits.

When a high enough voltage is applied across a distance occupied by a gas or vapor, or when that gas or vapor is heated sufficiently, the electrons of those gas molecules will be stripped away from their respective nuclei, creating a condition of *ionization*. Having freed the electrons from their electrostatic bonds to the atoms' nuclei, they are free to migrate in the form of a current, making the ionized gas a relatively good conductor of electricity. In this state, the gas is more properly referred to as a *plasma*.

Ionized gas is not a perfect conductor. As such, the flow of electrons through ionized gas will tend to dissipate energy in the form of heat, thereby helping to keep the gas in a state of ionization.

The result of this is a tube that will begin to conduct under certain conditions, then tend to stay in a state of conduction until the applied voltage across the gas and/or the heat-generating current drops to a minimum level.

The astute observer will note that this is precisely the kind of behavior exhibited by a class of semiconductor devices called "thyristors," which tend to stay "on" once turned "on" and tend to stay "off" once turned "off." Gas-filled tubes, it can be said, manifest this same property of *hysteresis*.

Unlike their vacuum counterparts, ionization tubes were often manufactured with no filament (heater) at all. These were called *cold-cathode* tubes, with the heated versions designated as *hot-cathode* tubes. Whether or not the tube contained a source of heat obviously impacted the characteristics of a gas-filled tube, but not to the extent that lack of heat would impact the performance of a hard-vacuum tube.

The simplest type of ionization device is not necessarily a tube at all; rather, it is constructed of two electrodes separated by a gas-filled gap. Simply called a *spark gap*, the gap between the electrodes may be occupied by ambient air, other times a special gas, in which case the device must have a sealed envelope of some kind.

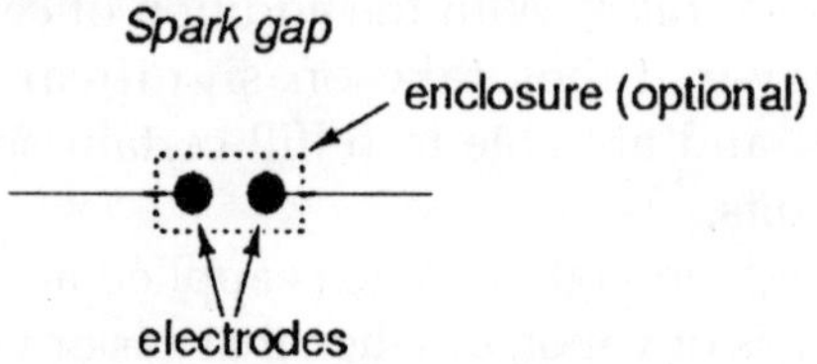

A prime application for spark gaps is in overvoltage protection. Engineered not to ionize, or "break down" (begin conducting), with normal system voltage applied across the electrodes, the spark gap's function is to conduct in the event of a significant increase in voltage.

Once conducting, it will act as a heavy load, holding the system voltage down through its large current draw and subsequent voltage drop along conductors and other series impedances.

In a properly engineered system, the spark gap will stop conducting ("extinguish") when the system voltage decreases to a normal level, well below the voltage required to initiate conduction.

One major caveat of spark gaps is their significantly finite life. The discharge generated by such a device can be quite violent, and as such will tend to deteriorate the surfaces of the electrodes through pitting and/or melting.

Spark gaps can be made to conduct on command by placing a third electrode (usually with a sharp edge or point) between the other two and applying a high voltage pulse between that electrode and one of the other electrodes.

The pulse will create a small spark between the two electrodes, ionizing part of the pathway between the two large electrodes, and enabling conduction between them if the applied voltage is high enough:

Triggered spark gap

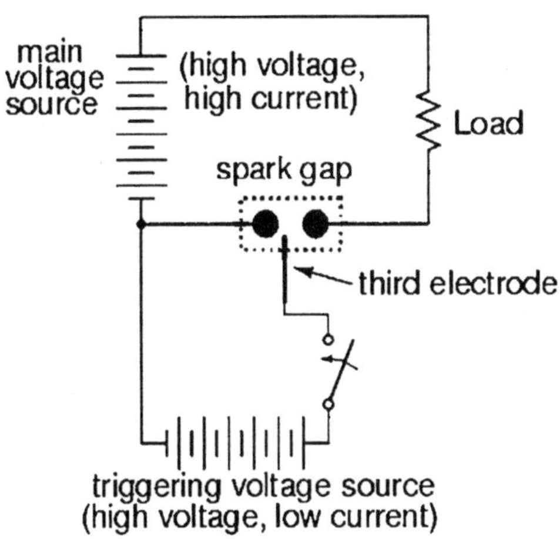

Spark gaps of both the triggered and untriggered variety can be built to handle huge amounts of current, some even into the range of mega-amps (millions of amps)! Physical size is the primary limiting factor to the amount of current a spark gap can safely and reliably handle.

When the two main electrodes are placed in a sealed tube filled with a special gas, a *discharge tube* is formed. The most common type of discharge tube is the neon light, used popularly as a source of colorful illumination, the color of the light emitted being dependent on the type of gas filling the tube.

Construction of neon lamps closely resembles that of spark gaps, but the operational characteristics are quite different:

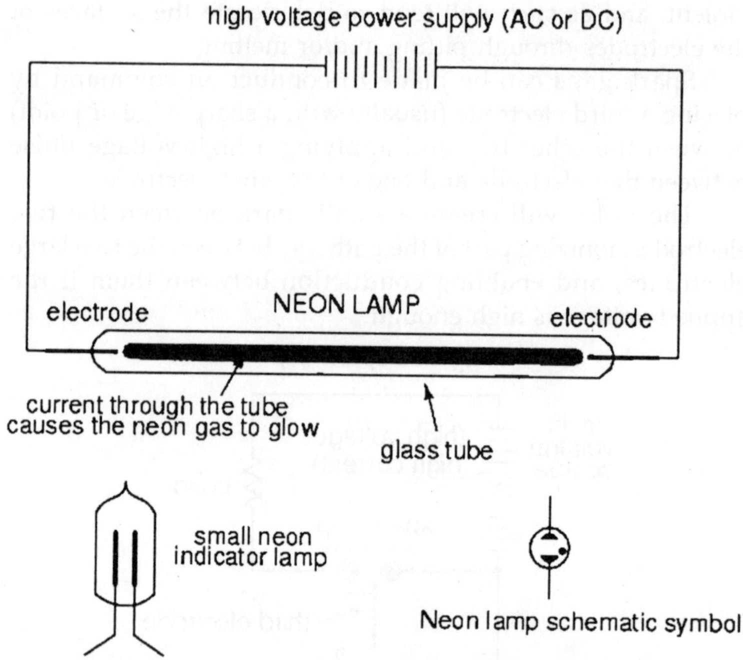

By controlling the spacing of the electrodes and the type of gas in the tube, neon lights can be made to conduct without drawing the excessive currents that spark gaps do.

They still exhibit hysteresis in that it takes a higher voltage to initiate conduction than it does to make them "extinguish," and their resistance is definitely nonlinear (the more voltage applied across.

the tube, the more current, thus more heat, thus lower resistance). Given this nonlinear tendency, the voltage across a neon tube must not be allowed to exceed a certain limit, lest the tube be damaged by excessive temperatures.

This nonlinear tendency gives the neon tube an application other than colorful illumination: it can act somewhat like a zener diode, "clamping" the voltage across it by drawing more and more current if the voltage decreases. When used in this fashion, the tube is known as a *glow tube*, or *voltage-regulator tube*, and was a popular means of voltage regulation in the days of electron tube circuit design.

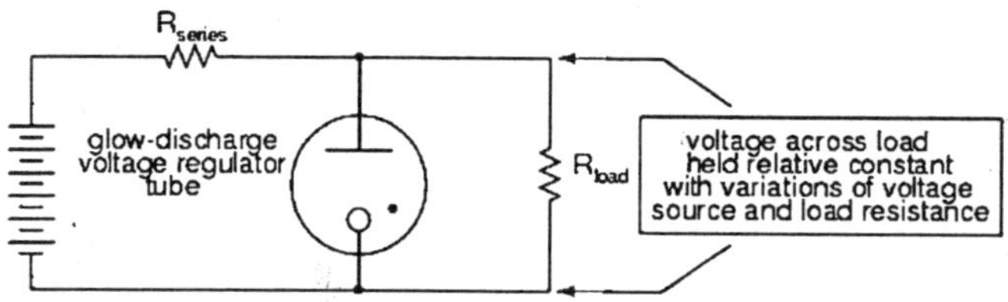

Please take note of the black dot found in the tube symbol shown above (and in the neon lamp symbol shown before that). That marker indicates the tube is gas-filled. It is a common marker used in all gas-filled tube symbols.

One example of a glow tube designed for voltage regulation was the VR-150, with a nominal regulating voltage of 150 volts. Its resistance throughout the allowable limits of current could vary from 5 kΩ to 30 kΩ, a 6:1 span. Like zener diode regulator circuits of today, glow tube regulators could be coupled to amplifying tubes for better voltage regulation and higher load current ranges.

If a regular triode was filled with gas instead of a hard vacuum, it would manifest all the hysteresis and nonlinearity of other gas tubes with one major advantage: the amount of voltage applied between grid and cathode would determine the minimum plate-to cathode voltage necessary to initiate conduction. In essence, this tube was the equivalent of the semiconductor SCR (Silicon-Controlled Rectifier), and was called the *thyratron*.

(Simple) Thyratron control circuit

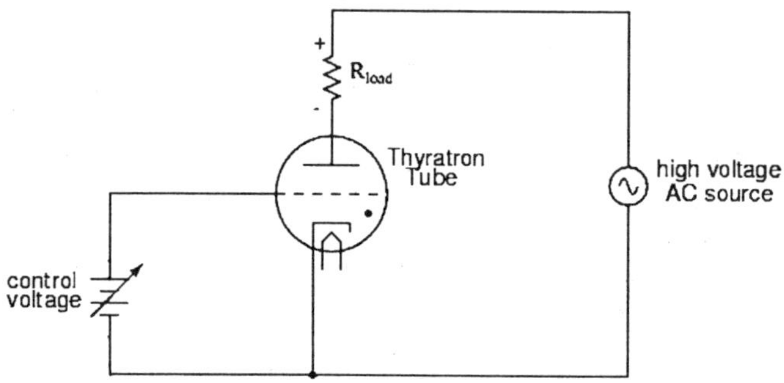

It should be noted that the schematic shown above is greatly simplified for most purposes and thyratron tube designs. Some thyratrons, for instance, required that the grid voltage switch polarity between their "on" and "off" states in order to properly work. Also, some thyratrons had more than one grid!

Thyratrons found use in much the same way as SCR's find use today: controlling rectified AC to large loads such as motors. Thyratron tubes have been manufactured with different types of gas fillings for different characteristics: inert (chemically non-reactive) gas, hydrogen gas, and mercury (vaporized into a gas form when activated). Deuterium, a rare isotope of hydrogen, was used in some special applications requiring the switching of high voltages.

DISPLAY TUBES

In addition to performing tasks of amplification and switching, tubes can be designed to serve as display devices.

Perhaps the best-known display tube is the *cathode ray tube*, or *CRT*. Originally invented as an instrument to study the behavior of "cathode rays" (electrons) in a vacuum, these tubes developed into instruments useful in detecting voltage, then later as video projection devices with the advent of television. The main difference between CRTs used in oscilloscopes and CRTs used in televisions is that the oscilloscope variety exclusively use electrostatic (plate) deflection, while televisions use electromagnetic (coil) deflection. Plates function much better than coils over a wider range of signal frequencies, which is great for oscilloscopes but irrelevant for televisions, since a television electron beam sweeps vertically and horizontally at fixed frequencies. Electromagnetic deflection coils are much preferred in television CRT construction because they do not have to penetrate the glass envelope of the tube, thus decreasing the production costs and increasing tube reliability.

An interesting "cousin" to the CRT is the *Cat-Eye* or *Magic-Eye* indicator tube. Essentially, this tube is a voltage-measuring device with a display resembling a glowing green ring. Electrons emitted by the cathode of this tube impinge on a fluorescent screen, causing the green-colored light to be emitted.

The shape of the glow produced by the fluorescent screen varies as the amount of voltage applied to a grid changes:

"Cat-Eye" indicator tube displays

The width of the shadow is directly determined by the potential difference between the control electrode and the fluorescent screen. The control electrode is a narrow rod placed between the cathode and the fluorescent screen. If that control electrode (rod) is significantly more negative than the fluorescent screen, it will deflect some electrons away from the that area of the screen.

The area of the screen "shadowed" by the control electrode will appear darker when there is a significant voltage difference between the two. When the control electrode and fluorescent screen are at equal potential (zero voltage between them), the shadowing effect will be minimal and the screen will be equally illuminated.

The schematic symbol for a "cat-eye" tube looks something like this:

"Cat-Eye" or "Magic-Eye" indicator tube

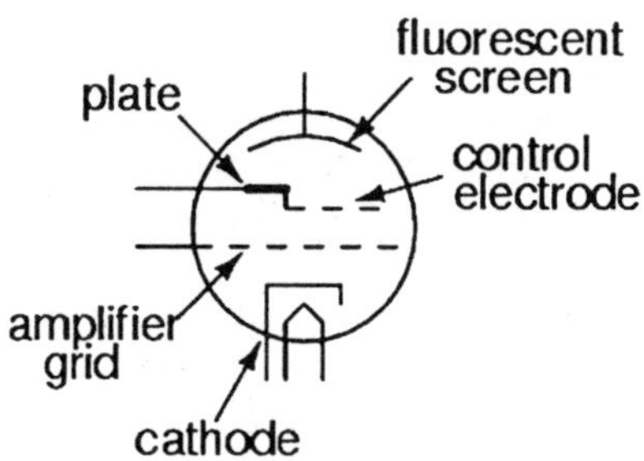

Here is a photograph of a cat-eye tube, showing the circular display region as well as the glass envelope, socket (black, at far end of tube), and some of its internal structure:

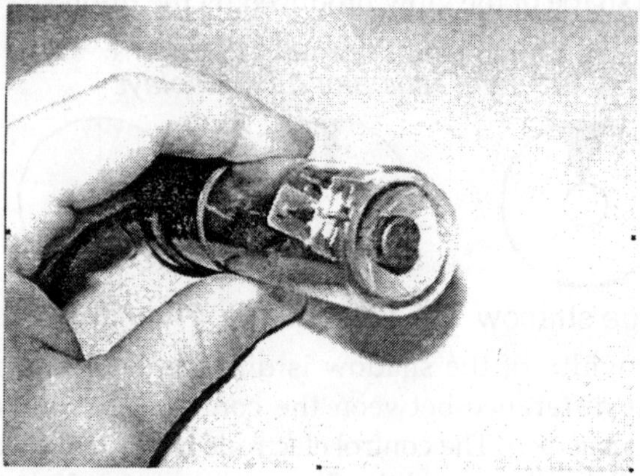

Normally, only the end of the tube would protrude from a hole in an instrument panel, so the user could view the circular, fluorescent screen.

In its simplest usage, a "cat-eye" tube could be operated without the use of the amplifier grid. However, in order to make it more sensitive, the amplifier grid *is* used, and it is used like this:

"Cat-Eye" indicator tube circuit

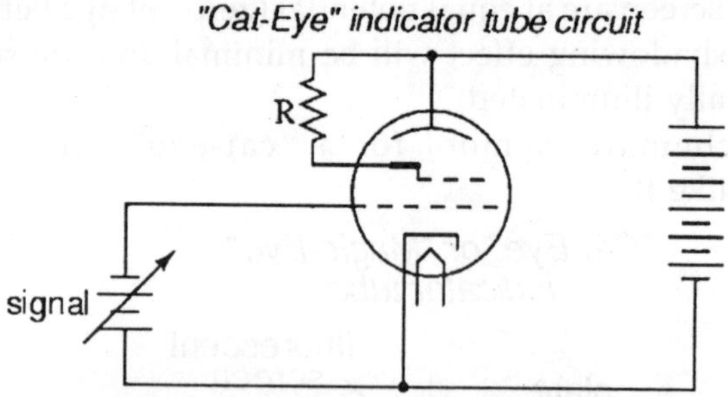

As the signal voltage increases, current through the tube is choked off. This decreases the voltage between the plate and the fluorescent screen, lessening the shadow effect (shadow narrows).

The cathode, amplifier grid, and plate act as a triode to create large changes in plate-to-cathode voltage for small changes in grid-to-cathode voltage.

Because the control electrode is internally connected to the plate, it is electrically common to it and therefore possesses the same amount of voltage with respect to the cathode that the plate does. Thus, the large voltage changes induced on the plate due to small voltage changes on the amplifier grid end up causing large changes in the width of the shadow seen by whoever is viewing the tube.

Control electrode negative with respect to the fluorescent screen. This is caused by a positive amplifier grid voltage (with respect to the cathode).

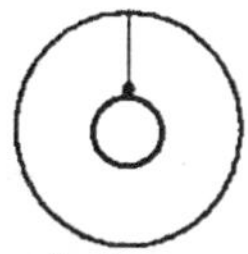

No voltage between control electrode and flourescent screen. This is caused by a negative amplifier grid voltage (with respect to the cathode).

"Cat-eye" tubes were never accurate enough to be equipped with a graduated scale as is the case with CRT's and electromechanical meter movements, but they served well as null detectors in bridge circuits, and as signal strength indicators in radio tuning circuits. An unfortunate limitation to the "cat-eye" tube as a null detector was the fact that it was not directly capable of voltage indication in both polarities.

MICROWAVE TUBES

For extremely high-frequency applications (above 1 GHz), the interelectrode capacitances and transit-time delays of standard electron tube construction become prohibitive. However, there seems to be no end to the creative ways in which tubes may be constructed, and several high-frequency electron tube designs have been made to overcome these challenges.

It was discovered in 1939 that a toroidal cavity made of conductive material called a *cavity resonator* surrounding an electron beam of oscillating intensity could extract power from the beam without actually intercepting the beam itself. The oscillating electric and magnetic fields associated with the beam "echoed" inside the cavity, in a manner similar to the sounds of traveling automobiles echoing in a roadside canyon, allowing radio-frequency energy to be transferred from the beam to a waveguide or coaxial cable connected to the resonator with a coupling loop. The tube was called an *inductive output tube*, or *IOT*:

The inductive output tube (IOT)

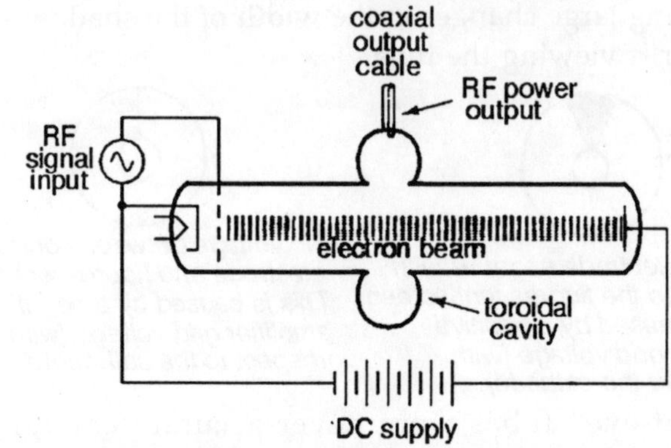

Two of the researchers instrumental in the initial development of the IOT, a pair of brothers named Sigurd and Russell Varian, added a second cavity resonator for signal input to the inductive output tube.

This input resonator acted as a pair of inductive grids to alternately "bunch" and release packets of electrons down the drift space of the tube, so the electron beam would be composed of electrons traveling at different velocities. This "velocity modulation" of the beam translated into the same sort of amplitude variation at the output resonator, where energy was extracted from the beam. The Varian brothers called their invention a *klystron*.

The klystron tube

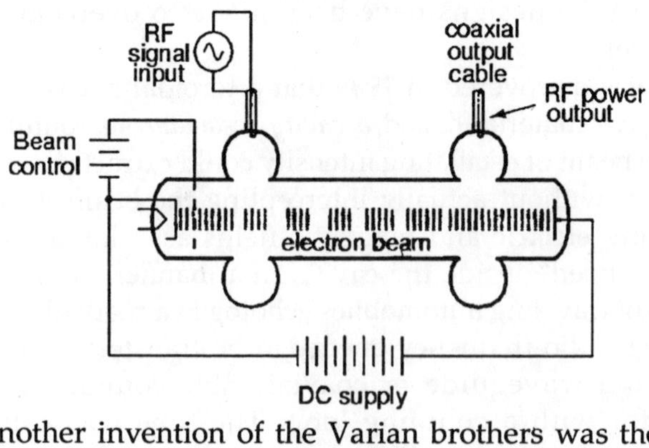

Another invention of the Varian brothers was the *reflex klystron* tube. In this tube, electrons emitted from the heated cathode travel through the cavity grids toward the repeller plate, then are repelled and returned back the way they came (hence the name *reflex*) through the cavity grids. Self-sustaining oscillations would develop in this tube, the frequency of which could be changed by adjusting the repeller voltage. Hence, this tube operated as a voltage-controlled oscillator.

The reflex klystron tube

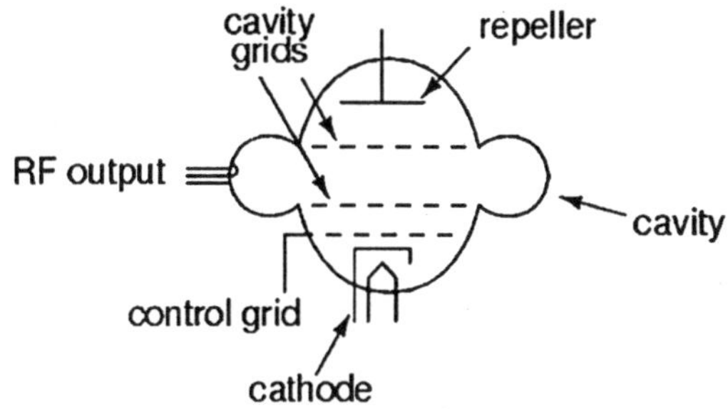

As a voltage-controlled oscillator, reflex klystron tubes served commonly as "local oscillators" for radar equipment and microwave receivers:

Reflex klystron tube used as a voltage-controlled oscillator

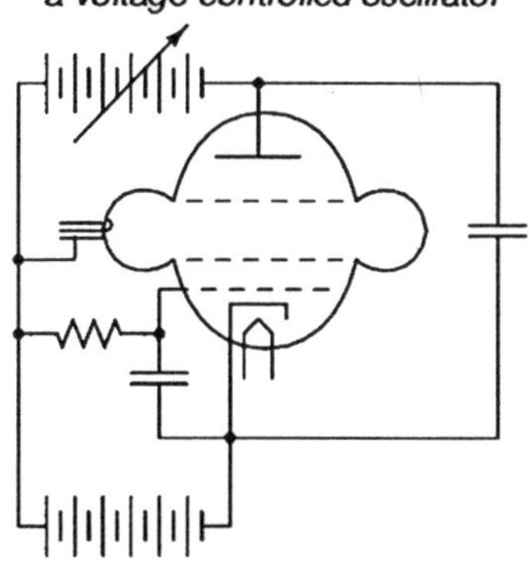

Initially developed as low-power devices whose output required further amplification for radio transmitter use, reflex klystron design was refined to the point where the tubes could serve as power devices in their own right. Reflex klystrons have since been superseded by semiconductor devices in the application of local oscillators, but amplification klystrons continue to find use in high-power, high-frequency radio transmitters and in scientific research applications.

One microwave tube performs its task so well and so cost-effectively that it continues to reign supreme in the competitive realm of consumer electronics: the magnetron tube. This device forms the heart of every microwave oven, generating several hundred watts of microwave RF energy used to heat food and beverages, and doing so under the most grueling conditions for a tube: powered on and off at random times and for random durations.

Magnetron tubes are representative of an entirely different kind of tube than the IOT and klystron. Whereas the latter tubes use a linear electron beam, the magnetron directs its electron beam in a circular pattern by means of a strong magnetic field:

The magnetron tube

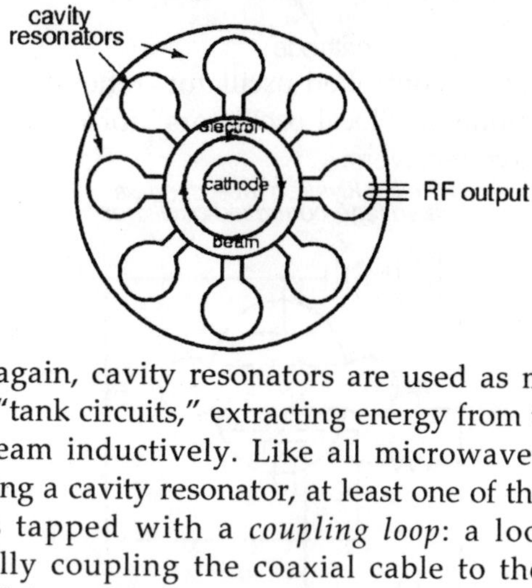

Once again, cavity resonators are used as microwave-frequency "tank circuits," extracting energy from the passing electron beam inductively. Like all microwave-frequency devices using a cavity resonator, at least one of the resonator cavities is tapped with a *coupling loop*: a loop of wire magnetically coupling the coaxial cable to the resonant structure of the cavity, allowing RF power to be directed out of the tube to a load. In the case of the microwave oven, the output power is directed through a waveguide to the food or drink to be heated, the water molecules within acting as tiny load resistors, dissipating the electrical energy in the form of heat.

The magnet required for magnetron operation is not shown in the diagram. Magnetic flux runs perpendicular to the plane of the circular electron path. In other words, from the view of the tube shown in the diagram, you are looking straight at one of the magnetic poles.

TUBES VERSUS SEMICONDUCTORS

Devoting a whole chapter in a modern electronics text to the design and function of electron tubes may seem a bit strange, seeing as how semiconductor technology has all but obsoleted tubes in almost every application. However, there is merit in exploring tubes not just for historical purposes, but also for those niche applications that necessitate the qualifying phrase "*almost* every application" in regard to semiconductor supremacy.

In some applications, electron tubes not only continue to see practical use, but perform their respective tasks better than any solid-state device yet invented. In some cases the performance and reliability of electron tube technology is *far* superior.

In the fields of high-power, high-speed circuit switching, specialized tubes such as hydrogen thyratrons and krytrons are able to switch far larger amounts of current, far faster than any semiconductor device designed to date. The thermal and temporal limits of semiconductor physics place limitations on switching ability that tubes — which do not operate on the same principles — are exempt from.

In high-power microwave transmitter applications, the excellent thermal tolerance of tubes alone secures their dominance over semiconductors. Electron conduction through semiconducting materials is greatly impacted by temperature. Electron conduction through a vacuum is not. As a consequence, the practical thermal limits of semiconductor devices are rather low compared to that of tubes. Being able to operate tubes at far greater temperatures than equivalent semiconductor devices allows tubes to dissipate more thermal energy for a given amount of dissipation area, which makes them smaller and lighter in continuous high power applications.

Another decided advantage of tubes over semiconductor components in high-power applications is their rebuildability.

When a large tube fails, it may be disassembled and repaired at far lower cost than the purchase price of a new tube. When a semiconductor component fails, large or small, there is generally no means of repair.

The following photograph shows the front panel of a 1960's vintage 5 kW AM radio transmitter. One of two "Eimac" brand power tubes can be seen in a recessed area, behind the glass door. According to the station engineer who gave the facility tour, the rebuild cost for such a tube is only $800: quite inexpensive compared to the cost of a new tube, and still quite reasonable in contrast to the price of a new, comparable semiconductor component.

Tubes, being less complex in their manufacture than semiconductor components, are potentially cheaper to produce as well, although the huge volume of semiconductor device production in the world greatly offsets this theoretical advantage. Semiconductor manufacture is quite complex, involving many dangerous chemical substances and necessitating super-clean assembly environments. Tubes are essentially nothing more than glass and metal, with a vacuum seal. Physical tolerances are "loose" enough to permit hand-assembly of vacuum tubes, and the assembly work need not be done in a "clean room" environment as is necessary for semiconductor manufacture.

One modern area where electron tubes enjoy supremacy over semiconductor components is in the professional and high-end audio amplifier markets, although this is partially due to musical culture. Many professional guitar players, for example, prefer tube amplifiers over transistor amplifiers because of the specific distortion produced by tube circuits. An electric guitar amplifier is designed to *produce distortion* rather than avoid distortion as is the case with audio-reproduction amplifiers (this is why an electric guitar sounds so much different than an acoustical guitar), and the type of distortion produced by an amplifier is as much a matter of personal taste as it is technical measurement. Since rock music in particular was born with guitarists playing tube-amplifier equipment, there is a significant level of "tube appeal" inherent to the genre itself, and this appeal shows itself in the continuing demand for "tubed" guitar amplifiers among rock guitarists.

As an illustration of the attitude among some guitarists, consider the following quote taken from the technical glossary page of a tube-amplifier website which will remain nameless:

Solid State: *A component that has been specifically designed to make a guitar amplifier sound bad. Compared to tubes, these devices can have a very long lifespan, which guarantees that your amplifier will retain its thin, lifeless, and buzzy sound for a long time to come.*

In the area of audio reproduction amplifiers (music studio amplifiers and home entertainment amplifiers), it is best for an amplifier to reproduce the musical signal with as *little* distortion as possible. Paradoxically, in contrast to the guitar amplifier market where distortion is a design goal, high-end audio is another area where tube amplifiers enjoy continuing consumer demand. Though one might suppose the objective, technical requirement of low distortion would eliminate any subjective bias on the part of audiophiles, one would be very wrong. The market for high-end "tubed" amplifier equipment is quite volatile, changing rapidly with trends and fads, driven by highly subjective claims of "magical" sound from audio system reviewers and salespeople. As in the electric guitar world, there is no small measure of cult-like devotion to tube amplifiers among some quarters of the audiophile world. As an example of this irrationality, consider the design of many ultra-high-end amplifiers, with chassis built to display the working tubes openly, even though this physical exposure of the tubes obviously enhances the undesirable effect of *microphonics* (changes in tube performance as a result of sound waves vibrating the tube structure).

Having said this, though, there is a wealth of technical literature contrasting tubes against semiconductors for audio power amplifier use, especially in the area of distortion analysis. More than a few competent electrical engineers prefer tube amplifier designs over transistors, and are able to produce experimental evidence in support of their choice. The primary difficulty in quantifying audio system performance is the uncertain response of human hearing. *All* amplifiers distort their input signal to some degree, especially when overloaded, so the question is which type of amplifier design distorts the least.

However, since human hearing is very nonlinear, people do not interpret all types of acoustic distortion equally, and so some amplifiers will sound "better" than others even if a quantitative distortion analysis with electronic instruments indicates similar distortion levels. To determine what type of audio amplifier will distort a musical signal "the least," we must regard the human ear and brain as part of the whole acoustical system. Since no complete model yet exists for human auditory response, objective assessment is difficult at best. However, some research indicates that the characteristic distortion of tube amplifier circuits (especially when overloaded) is less objectionable than distortion produced by transistors.

Tubes also possess the distinct advantage of low "drift" over a wide range of operating conditions. Unlike semiconductor components, whose barrier voltages, β ratios, bulk resistances, and junction capacitances may change substantially with changes in device temperature and/or other operating conditions, the fundamental characteristics of a vacuum tube remain nearly constant over a wide range in operating conditions, because those characteristics are determined primarily by the physical dimensions of the tube's structural elements (cathode, grid(s), and plate) rather than the interactions of subatomic particles in a crystalline lattice.

This is one of the major reasons solid-state amplifier designers typically engineer their circuits to maximize power-efficiency even when it compromises distortion performance, because a power-inefficient amplifier dissipates a lot of energy in the form of waste heat, and transistor characteristics tend to change substantially with temperature.

Temperature-induced "drift" makes it difficult to stabilize "Q" points and other important performance-related measures in an amplifier circuit. Unfortunately, power efficiency and low distortion seem to be mutually exclusive design goals.

For example, class A audio amplifier circuits typically exhibit very low distortion levels, but are very wasteful of power, meaning that it would be difficult to engineer a solid-state class A amplifier of any substantial power rating due to the consequent drift of transistor characteristics. Thus, most solid-state audio amplifier designers choose class B circuit configurations for greater efficiency, even though class B designs are notorious for producing a type of distortion known as *crossover distortion*. However, with tubes it is easy to design a stable class A audio amplifier circuit because tubes are not as adversely affected by the changes in temperature experienced in a such a power-inefficient circuit configuration.

Tube performance parameters, though, tend to "drift" more than semiconductor devices when measured over long periods of time (years). One major mechanism of tube "aging" appears to be vacuum leaks: when air enters the inside of a vacuum tube, its electrical characteristics become irreversibly altered. This same phenomenon is a major cause of tube mortality, or why tubes typically do not last as long as their respective solid-state counterparts. When tube vacuum is maintained at a high level, though, excellent performance and life is possible. An example of this is a klystron tube (used to produce the high-frequency radio waves used in a radar system) that lasted for 240,000 hours of operation .

If nothing else, the tension between audiophiles over tubes versus semiconductors has spurred a remarkable degree of experimentation and technical innovation, serving as an excellent resource for those wishing to educate themselves on amplifier theory. Taking a wider view, the versatility of electron tube technology (different physical configurations, multiple control grids) hints at the potential for circuit designs of far greater variety than is possible using semiconductors. For this and other reasons, electron tubes will never be "obsolete," but will continue to serve in niche roles, and to foster innovation for those electronics engineers, inventors, and hobbyists who are unwilling to let their minds by stifled by convention.

Index